Nuclear Energy

Nuclear Energy

R.K. Taneja

CYBER TECH PUBLICATIONS
4264/3, Ansari Road, Daryaganj, New Delhi-110002 (India)
Ph.: 011-23244078. Fax: 011-23280028
E-Mail: cyberpublicationsdelhi@yahoo.com

NUCLEAR ENERGY

R.K. Taneja

First Edition 2009

Published by :

G.S. Rawat for **Cyber Tech Publications**
4264/3, Ansari Road, Daryaganj, New Delhi-110002 (India)
Ph.: 011-23244078. Fax: 011-23280028
E-Mail: cyberpublicationsdelhi@yahoo.com

Printed at Asian Offset Printers

PREFACE

This book covers the course in Nuclear Energy for most of universities in India. The language of the book has been kept as simple as could be consistent with precision and brevity.

The text have on purpose been kept fast, partly to reduce the volume of the book, and partly to allow for different approaches by the individual students. Further, the emphasis must be on the physical assumptions underlying mathematical procedures of integration, summation etc. The problems from an integral part of the text. It is how the standard recognized practice in all advanced teaching methods the problems must not be limited to reproduction of some parts of the text, but must lead to projection of the students mind beyond the subject matter of the text, recent examination trends are also in that direction, there fore, we have not simply reproduced examination questions from different Universities but have built up a structure of problems which have a great teaching value in themselves.

Contents

Pages

Preface

1. **Nuclear Structure** 1

Electronic Structure in Atoms, Fine Structure, X-Ray Spectra, Experimental Confirmation of the Vector Model Stern and Gerlach Experiment, Interpretation of Results, Wave Mechanical Atom Model, Some Examples of Electron Configurations, Fine Structure of Spectral Lines, Optical Spectra, Solved Examples, Exercises.

2. **Nuclear Emulsion** 66

Introduction, Pulse Amplifier, Detectors of Solid-state, Regions of Multiplicative Operation, Proportional Counter, Cerenkov Counters, Cloud Chamber, Spark Chamber, Chamber of Ionization, Bubble Chamber, Geiger-müller Counter, Scintillation Counters, Solved Examples, Exercises.

3. **Nuclear Transmutation and Reactions** 129

Introduction, Continuum Theory of Nuclear Reaction, Resonance : Breit Wigner Dispersion Formula, Optical Model, Different Stages of a Nuclear Reaction, Statistical Theory of Nuclear Reactions, Direct Reactions, Stripping Reactions and the shell model, Inverse Process, Cross-section of a Nuclear Reaction, Nuclear Reactions, Conservation Laws, Classical Analysis of Cross-section, Partial-wave Analysis of Reaction Cross-section, Compound Nucleus, Nuclear Reaction Kinematics, Charged Particle Reaction Spectroscopy, Neutron Spectroscopy, Nuclear Cross Section, Solved Examples, Exercises.

4. **Nuclear Detection** 191

Introduction, Slowing Down Neutrons, Energy Distribution of thermal Neutrons, Neutron Velocity Selector, Discovery of Neuton, Neutron Sources, Neutron Collimators, Basic Properties of the Neutron, Classification of Neutrons as to Energy, Neutron Diffusion, Solved Examples, Exercises.

5. **Nuclear Reactors** 241

Nuclear Chain Reaction, The Critical Size of a Reactor, General Aspect of Reactor Design, Classification of Reactors, Solved Examples, Exercises.

1

Nuclear Structure

ELECTRONIC STRUCTURE IN ATOMS

Admitting that the number of electrons in the normal atom increases steadily as one passes from light to heavy elements, one is interested to know whether the several electrons in the given atom are distributed at random or in a regular manner. Since, in general, a system is most stable when its energy is minimum, and as the energy of the atom is smaller, the lower the value of the total quantum number n designating the electronic orbit, it would seem that all the electrons should tend to crowd into the lower orbit of least energy. But investigations based on X-ray spectra and the periodic classification of elements have proved the contrary *viz.,* the electrons, instead of all crowding into the lowest orbit, are arranged in different orbits with a certain regularity.

The study of X-ray spectra of the elements strongly suggests that the *electrons in the atom are arranged in different groups or shells* round the nucleus, all the electrons having the same total quantum number forming one shell. Thus the innermost shell, known as the **K** shell, consists of electrons whose total quantum number n = 1; the next shell, called the **L** shell, contains electrons for which n = 2 and so on. The complete scheme, as far as it is required in building up the electronic structure in the atoms of all the elements is:

Shell	→	K	L	M	N	O	P	Q
n	→	1	2	3	4	5	6	7

For the conception and nomenclature of these electronic shells, we are indebted to Lewis, Langmuir, Bohr and Stoner. According to the conception of orbits, as there may be several orbits having the same total quantum number, all the orbits having the same value of n are said to form one shell.

The *periodic table of elements,* with its clear manifestation of periodic variations in the properties of the elements, confirm, in its turn, such a grouping of electrons in different shells. The periodic table is an arrangement of the different elements found in nature in a scheme, based on their chemical properties and atomic weights, first drawn up by Mendeleeff in 1871 and then gradually perfected, chiefly by the study of X-ray spectra of the elements. Moseley's work on the characteristic X-ray lines, for instance, settled in an unambiguous manner that the determining factor in the arrangement is not the atomic weight but the atomic number which assigns the numerical position of the element in the classification, as l, 2, 3..., up to 92. This table of elements presents a complex scheme, with *seven* horizontal rows called *periods,* in which the chemical and physical properties of the elements vary gradually as a periodic function of the atomic number, and eight vertical columns, called *groups,* each of which contains elements manifesting similar properties.

The number of elements in the different periods is not the same; the first consists of only two elements, the second and third *eight* elements each, the fourth and fifth eighteen elements each, the sixth thirty-two and the seventh sixteen elements. As the numerical position of the elements is determined by the atomic number, it gives also the number of electrons in their respective atoms.

Such a periodic classification of elements readily suggests the idea of the *formation, at regular intervals, of a new shell*, which, as it develops in the course of the period, passes through the same characteristic phases, determined by the number and distribution of its component electrons. There is also a further implication that a new shell which marks the beginning of a new period *will*, in general, be formed *only when the inner shells are completed* with their due quota of electrons. It must, however, be noted that there occur cases where a new outer shell is started before the inner one is filled.

To determine the *number of electrons required to complete each one of the shells*, we have a clue in the periodic occurrence of monotonic elements chemically inactive, hence known as inert gasses, *viz.*, helium, neon, argon, krypton, zenon and radon (classified under group 0 in the table). Considering their atomic numbers, *viz.*, 2, 10, 18, 36, 54, and 86 respectively, the total number of electrons in their respective atoms are also 2,10, 18, 36, 54 and 86. Rydberg made a suggestion that these numbers could be arranged in a simple numerical series as follows: Such an arrangement indicates *that electrons in these inert gases can be distributed in shells*, that the helium atom, for instance, has a single shell of two electrons, neon has two shells, the first containing two electrons and the second eight electrons, argon has three shells, the first one containing two electrons, the second eight, the third eight also, krypton has four shells, the first three containing the same number of electrons as argon, but the fourth eighteen electrons and so on.

The factor two outside the bracket suggests some sort of symmetry, pairing of electrons. Further investigations showed that Rydberg's series had to be modified as regards the order of the numbers within the bracket. Thus, in the case of Kr the arrangement should be $2(1^2 + 2^2 + 3^2 + 2^2)$, the outermost shell containing eight electrons like A, but the third eighteen electrons. Similarly, for Xe the order should be $2(1^2 + 2^2 + 3^2 + 2^2)$ and for Rn $2(1^2 + 2^2 + 3^2 + 4^2 + 3^2 + 2^2)$. This *would indicate that the number of electrons in the shells increases up to a certain limit and then decreases.*

The chemical inertness of these elements shows that they atoms have very stable structure, with a peculiar permanent nature for their electrons. This, in turn, suggests the *general law that* 2, 10, 18, 36, etc. *electrons tend to make a very stable combination in the atom.* On the other hand, the next element beyond any one of these inert gases has one more electron than necessary to form a stable structure, *e.g.*, Li (3), Na (11), K (19), Rb (37), Cs (55), the metals of the alkali group, whose properties are strikingly similar. The second element beyond an inert gas has two superfluous electrons, *e.g.*, Be (4), Mg (12), Ca (20), Sr (38), Ba (56), Ra (88), the alkaline earths, also with similar properties. Going in the other direction, the next element before inert gas lacks one electron to make a stable structure, *e.g.*, F (9), Cl (17), Br (35), I (53), forming the halogen, group, again with similar properties. The second next before the inert gases, *viz.* 0 (8), S (16). Se (34), Te (52) are likewise similar. Considerations of this kind indicate *a very fundamental principle in electronic structure, viz., the chemical and physical properties of elements are determined by the number and arrangement of the electrons in the outermost shell rather than by the total number of electrons in their respective atoms.*

Conversely, by studying the chemical and physical properties of elements one may obtain information as to the number of electrons in the outermost shell of their respective atoms. Then paying attention to the elements with respect to the nearest inert gas, it is possible to determine the whole arrangement of the electrons in the different shells. Thus for instance, in the case of Na the arrangement is two electrons in the first K shell, eight electrons in the second L shell and one in the third M shell as yet incomplete, represented ordinarily by the abbreviation (2, 8, 1). Similarly, the arrangement in Mg is (2, 8, 2) in F (2, 7) and in 0 (2, 6). Reasoning of this kind determine also the number of electrons required to complete the different shells. The K shell is complete or closed with 2 electrons, the N shell with 32 electrons, the 0 shell with 18, the P shell with 12 and the Q shell with 2.

Thus, the shell character of the electronic structure in atoms, which manifests itself in the periodic variation of the chemical and physical properties of elements and their consequent classification into a periodic system, was discovered by a close scrutiny of empirical data.

Pauli's Exclusion Principle gives a rational theory; of such an empirically deduced electronic shell structure of atoms, showing why the electrons are grouped in shells, why a certain definite number of them is required to complete a shell and why the number and distribution of electrons outside closed shells are mainly responsible for the chemical and physical properties of atom, thus, leading up to a *logical construction of the periodic table.*

Considering the **K** shell, for the electrons in it n = 1, l = 0; hence m_l = 0. Thus, three of the four quantum numbers that characterise an electron are fixed. The fourth, *viz.* m_s can assume only two values +1/2 and –1/2. Hence the maximum number of electrons in **K** shell can, be *only two*, since a greater number would mean that two more should necessarily be identical as regards all their four specific quantum numbers, which is against Pauli's exclusion principle. *The shell is therefore completed or closed with 2 electrons.*

For the electrons in the next **L** shell n = 2, l = 0 and 1. In the case n = 2, l = 0, and m = 0 and m ± 1/2; hence there can be only two electrons, according to Pauli's Principle. In the other case n = 2, l = 1, m_l can have (2l + 1) values, *i.e.,* three values 1, 0, –1; with each of these may be associated the two values for the four quantum members characterising an electron, which are not identical; hence in this case six electrons can co-exist. *The L shell with two sub-groups ($n = 2, l = 0$ and $n = 2, l = 1$) is therefore completed when it contains $2 + 6 = 8$ electrons.*

The electrons in the third M shell have n = 3, and $l = 0, 1, 2$. Three sub-groups can therefore be distinguished in this shell, *viz.* first n = 3, $l = 0$, second n = 3, $l = 1$, and third n = 3, $l = 2$. The first and second sub-groups are completed by two and six electrons respectively for reasons given above. The third is completed with $2(2l + 1)$ electrons *i.e.*, ten electrons, since $l = 2$. Hence, *the total number of electrons required to complete the M shell is 18.*

Arguing in a similar manner, it can be shown that fourth **N** shell has four groups, since n = 4 and $l = 0, 1, 2, 3$, the first three are completed by 2, 6 and 10 electrons respectively, while the fourth by $2(2l + 1) = 2(2 \times 3 + 1) = 14$ electrons. This brings up the total to *32 electrons to complete the n shell.*

From this discussion the following general Theory of Atomic Structure also can be drawn:

(i) *In the electronic configuration of an atom there can be only $2n^2$ electrons with the same total quantum number n.*

(ii) *In the shell there are n sub-groups and in each of these sub-groups designated by the orbital quantum number l, there can be only $2(2l + 1)$ electrons.*

Thus, Pauli's Principle in conjuction with the vector model supplies the reason why electrons of an atom do not all occupy the most stable shell closest to the nucleus but dispose themselves in different shells according to certain regulations.

A direct consequence of Pauli's Principle is that every closed shell or sub-shell is balanced with respect to angular momenta both as regards orbital and spin motions and hence contributes nothing to the total angular momentum of the atom as a whole. The proof of this is essentially contained in the statement that a closed shell or orbital quantum number l is built up of $2(2l + 1)$ electrons. Since the (7.1 + 1) different values of m_l and the two different values of w, are just used up in this way by the $2(2l + 1)$ electrons, there is nothing more of m_l or m_s which can contribute to the atom as a whole. In other words, for each closed sub-group as well as for the whole closed shell, $M_l = \Sigma m_l = 0$ and $M_s = \Sigma m_s = 0$. Thus, for instance, taking the L shell, for the first closed sub-group (n = 2, l = 0), w; = 0 and the vector sum of the two values of m_s is equal to zero, so that $\Sigma m_s = 0$. For the second closed sub-group (n = 2, l = 1), the vector sum of the three m_l values, *viz.* +1, 0, –1 is evidently zero, so that $\Sigma m_l = 0$; similarly the vector sum of the two m_l values is also zero.

If, therefore, M_L and M_s are zero for a closed- sub-group or shell, L = 0, S = 0 and J = L + S = Q. A closed sub-shell or shell cannot, in consequence, contribute anything to the atom as a whole. From this it follows that the total angular momentum of the atom is determined only by the electron outside closed shells. *This fact explains why the chemical and physical properties of elements depend only on the electronic configuration outside closed shells, i.e.,* on the electrons in the outermost incomplete shells, which are therefore known as the valency or optical electrons.

Having thus fixed the maximum number of electrons in each shell and sub-shell, it is possible to build up elements genetically and show that they can be arranged in a periodic system, starting from hydrogen and ending at uranium. The general procedure is to picture the nucleus of atomic number Z, initially stripped of all its electrons, as capturing one by one of the Z electrons necessary to make a neutral normal atom. The quantum numbers chosen for each added electron are such as to place the electron in a vacant shell, subjected to Pauli's principle and to minimum energy, the condition required for maximum stability. In deciding which shell of sub-shell the next captured electron will permanently occupy, spectroscopic and critical potential data are used.

When normal atoms of different elements are built like this, it is found that their chemical and spectroscopic behaviour are determined by the electron configuration of the outermost shell, as is to be expected from Pauli's Principle. Elements with outermost shells of similar electronic structure possess in a large measure equivalent chemical a id physical properties. This explains the occurrence of periods in the classification of elements. Thus the alkalis, to which belong Li, Na, K, Rb and Cs are characterised by a single electron in the outermost shell, the alkaline earths, Mg, Sr, Ba, and Ra, have two electrons outside closed shells, the halogens, F, Cl, Br, and I, lack one electron to make up a closed shell; the inert gases, Ne, A, Kr, Xe and Rn have the outermost shell closed with eight electrons.

Pauli's exclusion principle leads, therefore, to a periodic system agreeing closely with the one built on experimental data. It further represents an *ideal* system, much simpler than the real one of chemists and going deeper into the internal structure of the atom, which are of interest to the chemist and the spectroscopist in the optical *i.e.,* visible region, while the ideal system describing the interior of the atom is a great asset to the X-ray spectroscopist.

It is to be noted that the *ideal periodic system given by Pauli's exclusion principle does not however, completely agree with the real one.* The two coincide up to argon, but then with potassium and calcium, the N shell begins to form before the completion of the preceding M shell; likewise the 0 shell is initiated irregularly with Rb(37), while it has to begin with Hf(72).

Another discrepancy is that in the real system the 0 shell has only 18 electrons although 50 electrons would be required according to the ideal system to complete it, the P shell only 12 electrons, while 72 would be necessary to fill it completely and the Q shell 2 electrons alone when 98 would be its maximum quota. These differences between the real and ideal systems are accounted for as follows : The minimum energy condition may not always be such that one shell must be completed before an electron settles in the next. Further, since the deviations of the real system form the ideal one concern only the less tightly bound electrons of the outermost shells, if we regard as representatives of the elements not only the normal neutral atoms but also the ionised ones, the theoretical rule is found to be exactly verified.

FINE STRUCTURE

As spectral lines arise due to transitions of the atom from one spectral term to another, it is easy to understand, in a general way, how complex lines may appear when there are several spectral terms for a given atomic state defined by the vector L.

In the atoms belonging to the one-electron system all the states except the ground one are doublets. Transitions between these doublet states give rise to the fine structure. To account fully for the actually observed fine structure, as regards their number and intensity, the selection and intensity rules must be taken into consideration.

In the case of atoms pertaining to the many-electron system the fine structure may be expected to be much more complex on account of the greater term multiplicities involved. But here also, knowing the spectral terms and applying the selection and intensity rules, the observed fine structure can be accounted for. It is to be remarked that on account of the special property of closed shells the free electrons responsible for the spectra are limited to a small number. Hence the complexity of the fine structure does not go on increasing from one end to the other of the periodic table. The selection rules also, in their turn, put a limit on the complexity of fine structure. Likewise the limitations of term multiplicity arising from Pauli's exclusion principle may reduce the complexity still further.

These general principles may now be applied to the interpretation of the fine structure in the two simple cases already mentioned, *viz.*, the sodium D line and the hydrogen H_x line, to serve as illustrations.

Fine Structure of the Sodium D-Line

As we have already seen, ten out of the eleven electrons of the normal sodium atom are interlocked in closed shells, so that the eleventh free electron alone is responsible for the spectrum. The spectral lines arise from the various orbits which this single electron may occupy beyond the two closed shells and the state of the atom as a whole is determined by the state of that electron. For the same reason, all the states of the sodium atom except the ground one are doublets.

With the simple theory of Bohr it has been shown that the D-line belongs to the principal series which are due to transitions from the P state to the S state. Now for *the upper P state*, L = 1, J = 3/2 or 1/2 so that two terms are possible : $^2P_{3/2}$ and $^2P_{1/2}$. For the *lower S state*, L = 0, L = 1/2, so that only term is possible : $^2S_{1/2}$.

The transitions that can take place between the two terms of the P state and the single term of the S state are two:

$^2P_{1/2} \rightarrow {}^2S_{1/2}$ and $^2P_{3/2} \rightarrow {}^2S_{1/2}$

Now applying the selection rules $\Delta L = \pm 1$ and $\Delta J = 0$ or ± 1 (excluding $O \rightarrow 0$), both the transitions are allowed, which explains the doublet line structure of the sodium D-line. The D_1 component (5896Å) is due to the transition $P_{1/2} \rightarrow S_{1/2}$ and the D_2 component (5890 to $P_{3/2} \rightarrow$ Å) $S_{1/2}$.

According to the general intensity rules, D_2 will be more intense than D_1, since for D_2, $\Delta L = -1$, $\Delta J = -1$, while for D_1, $\Delta L = -1$, but $\Delta J = 0$.

Applying the total intensity (2J + 1) rule it can be shown that the ratio of the intensifies of D_2 to D_1 is 2 : 1, which has been fully confirmed by observation.

Note: *Fine structure of lines in the other series of sodium. As* the *sharp* series arises due to transitions between the S and P states, the lines in it will be *doublets*, as in the principal series.

The *diffuse* series arises due to transitions between the D and P states and hence the lines in it should be *triplets*, since both the states have two sub-terms each: $D_{5/2}$ and $D_{3/2}$, $P_{3/2}$, and $P_{1/2}$. Four transitions are therefore possible: $D_{5/2} \rightarrow P_{3/2}$, $D_{5/2} \rightarrow P_{1/2}$, $D_{3/2} \rightarrow P_{3/2}$ and $D_{3/2} \rightarrow P_{1/2}$. Of these, $D_{5/2} \rightarrow P_{1/2}$ is forbidden, since $\Delta J = 2$. The lines $D_{5/2} \rightarrow P_{3/2}$ and $D_{3/2} \rightarrow P_{3/2}$ will be intense, since $\Delta L = -1$ and $\Delta J = -1$, while $D_{3/2} \rightarrow P_{3/2}$ will be weak,, since $\Delta L = -1$ and $\Delta J = 0$. Experimentally, with ordinary resolving power instruments the weak line is not seen. When it was first discovered it was called a "satellite" of the apparent doublet formed by the two brighter lines.

Thus, the chief spectrum of the alkalis should consist of doublets and triplets and theory accounts for them very well.

Fine Structure of the H_α-Line

As the hydrogen atom, in whose spectrum the H_α-line appears, belongs to the one-electron system, L = I, S = s, and J = j and all the quantum states are doublets, except the singlet ground state. According to the simple theory of Bohr, the H_x-line the first of the Balmer Series, arises, due to transition from the third quantum state (n = 3) to the second (n = 2).

For the *upper state* (n = 3), L can take any of the values 0, 1, 2. Using the relation J = L ±S, since S in this case has the value 1/2, the possible number of terms are five, *viz.*, $^2D_{5/2}$, $^2D_{3/2}$, $^2P_{3/2}$, $^2P_{1/2}$ and $^2S_{1/2}$

For the lower state, (n = 2), L can have two values 0 and 1, which gives the possible number of terms as three, *viz.*, $^2P_{3/2}$,$^2P_{1/2}$ and $^2S_{1/2}$

Fifteen transitions are theoretically possible between the five terms of the upper state and the three terms of the lower state. But the selection rules $\Delta L = \pm 1$, and $\Delta J = 0$ or ± 1 reduce the permitted transitions to the following, seven:

$D_{5/2} \rightarrow P_{3/2}$, $D_{3/2} \rightarrow P_{3/2}$, $D_{3/2} \rightarrow P_{3/2}$, $D_{3/2} \rightarrow P_{1/2}$, $P_{3/2} \rightarrow S_{1/2}$

$P_{1/2} \rightarrow S_{3/2}$, $S_{1/2} \rightarrow P_{3/2}$ and $S_{1/2} \rightarrow P_{1/2}$

Two pairs in separate cases of these seven allowed transitions are identical *viz.*

$D_{3/2} \rightarrow P_{1/2}$ and $P_{3/2} \rightarrow S_{1/2}$, $P_{1/2} \rightarrow S_{1/2}$ and $S_{3/2} \rightarrow P_{1/2}$

since they represent transitions between *coincident levels*, *i.e.*, levels whose **L** values differ by units, but **J** values are the same.

Taking into account these identical lines, *the fine structure of the* H_α*-line should have five components.*

Applying the intensity rules, the two strongest components are:

$$D_{5/2} \rightarrow P_{3/2} \text{ and } \begin{Bmatrix} D_{3/2} & \rightarrow & P_{1/2} \\ P_{3/2} & \rightarrow & S_{1/2} \end{Bmatrix}$$

since they are of the type $\Delta L = -1$, $\Delta J = -1$.

The two components $D_{3/2} \rightarrow P_{3/2}$ and $P_{1/2} \rightarrow S_{1/2}$ are less intense being of the type $\Delta L = -1$, $\Delta J = 0$. The component $S_{1/2} \rightarrow P_{3/2}$ is, still weaker, since it is of the type $\Delta L = +1$, $\Delta J = +1$. The weakest line would be $S_{1/2} \rightarrow P_{1/2}$ ($\Delta L = +1$, $\Delta J = 0$), but it is fused with $P_{1/2} \rightarrow$ for reasons given above.

Energy level diagram (Fig. 1.28). The *coincident levels* are represented by horizontal lines drawn close together. There are two such levels in the upper state and one in the lower. The other lines represent, the *neighbouring* levels. If we call those vertical lines which terminate at the upper level of the lower state as group I and those which terminate at the lower level as group II and if we number the initial levels of the upper state as a, b and c, in no way as they do not coincide, then the two identical pairs are II_b, and II_c.

The five components are marked 1, 2, 3, 4 and 5, corresponding to I_a, I_b, I_c, II_b and II_c respectively. According to the intensity rules, the strongest components are I_a and II_b, (1 and 4). The components I_b and II_c (2 and 5) are less strong. I_c (3) is still weaker.

In practice, not all the five components but only doublets can be observed. This is due to the fact that the five components are merged together by the Doppler broadening caused by the thermal motion of the molecules. Very careful experiments conducted by R.C. Williams, in 1938, where the Doppler effect was reduced to a great extent by the use of a liquid air-cooled discharge tube containing deuterium, revealed three components.

X-RAY SPECTRA

Spectral terms. We have seen that each X-ray energy level represents a state of an atom which has one electron missing from a closed shell. Pauli pointed out that in a configuration, in which an electron is missing from a completed shell, the spectral term is the same as if that one electron occupied the shell. This means that the term scheme for X-ray spectra corresponds to the one-electron system, such as hydrogen, alkali metals, etc. Moseley's researches on the characteristic X-rays also suggest the same Theory of Atomic Structure, *viz.*, that the terms involved in X-ray spectra are of the hydrogen type.

The notations used are, however, different. The X-ray energy states are designated by the capital letters K, L, M, etc., specifying the shells to which they refer. According to the vector model, not only each of these states is characterised by a total quantum number n (being 1 for the K state, 2 for the L state, 3 for the M state, etc.), but also all the states, except K, are further split into sub-groups, L into two, M into three, N into four, etc.

For, using the fact that an atom emitting X-rays is assimilated to the one-electron system, it follows that in the mechanism of X-ray emission, L = 1, S = s, J = j, so that $J = j = l \pm 1/2$, and that *every distinct state of an atom except the lowest is a doublet*. Hence the *K state* (n = 1, l = 0) corresponding to the lowest ground state is a *singlet*. The *L state* (n = 2, l = 0, 1) have two sub-groups. The first sub-group (n = 2. l = 0) gives one term while the second sub-group (n = 2, l = 1) two terms corresponding to the two different values of s, *viz.*, 3/2 and 1/2. The *L state* thus, having three possible terms is a *triplet*. The *M state* (n = 3, I = 0, 1, 2 with three sub-groups gives rise to *quintet*, one corresponding to l = 0, and j = 1/2 (first sub-group), two others to l = 1, and j = 5/2, 3/2 (third sub-group). The *N state* (n = 4, l = 0, 1, 2, 3) with four sub-groups is a *septet* composed of a singlet for T = 0 and three *doublets* for l = 1, 2, 3. The term multiplicity does not, however, go on increasing. Pauli's principle putting a limit, the next *O state is a quintet and P state a triplet*. There are in all 24 different terms which are quite sufficient to represent completely the X-ray spectra of all the elements. They are denoted by the Roman numerals attached to the capital letters representing the states, *e.g.*,

$K, L_I\ L_{II}\ L_{III}, M_I\ M_{II}\ M_{III}\ M_{IV}\ M_V$, etc.

They indicate all the possible combinations of l and j for given values of n.

Fine Structure

The X-ray lines, in general,, will arise due to transitions among the 24 different X-ray terms. But by the selection rule transitions can only occur subject to the conditions $\Delta L = \pm 1$ and $\Delta J = 0$ or ± 1, so that the number of possible transitions between two states is thereby considerably reduced and in consequence the number of fine structure components of the X-ray lines will be limited.

If the transition ends in the K state the corresponding radiation is called a K line; if it ends in the L state, a L line and so on. Fine structure of these general classes are distinguished by small Greek letters in accordance with the following recognised convention : If the transition takes place between the different terms of the L state and the single term of the K state, the lines are called $K_{\alpha 1}$, $K_{\alpha 2}$, etc., if from the different terms of the M state to the K state, $K_{\beta 1}$, , $K_{\beta 2}$ etc. Thus a general classification is made as K, L, M, lines etc. according to the final state of transition, while the particular classification of the individual fine structure components of each line is made according to the terms of the initial state of transition.

These points are illustrated in Fig. 1.29. The term scheme of the four states K, L, M, N alone is drawn. The values of n, l and j for each term are indicated. Some of the fine structure of the K and L lines are marked. The continuous lines represent the permitted transitions giving rise to the fine structure components, while the dotted lines indicate transitions forbidden by the selection rules. As one proceeds to elements of higher atomic number, the number of terms involved and consequently the number of the fine structure lines will increase. This clear and simple interpretation of the fine structure of X-ray lines in a single scheme was obtained in 1927, thanks chiefly to the work of four physicists, Smekal, Coster and Bohr. Thus, the fine structure of spectral lines, both in the optical and X-ray regions, finds an adequate explanation on the basis of spectral terms determined by the vector model and of certain selection rules. It should be emphasized that the fine structure of spectral lines can be satisfactorily accounted for only by the introduction of electron spin.

EXPERIMENTAL CONFIRMATION OF THE VECTOR MODEL STERN AND GERLACH EXPERIMENT

Experimental study of the behaviour of atoms in non-homogeneous magnetic fields, initiated by Stern and Gerlach in 1921 and since then elaborated in many directions by different workers, gave a very direct and most convincing confirmation of the essential features of the vector model, *viz.*, spatial quantisation, spinning electron and quantised atomic magnetic moment.

Principle

The magnetic moments due to the orbital and spin motions of the electrons in an atom may be estimated as follows:

Considering the simple case of an electron moving in a central orbit, such as an ellipse, the current i due to the motion of the electron round the orbit is given by i = e/T, where e is the charge and T the periodic time.

Applying Ampere's Theorem, this current gives rise to a magnetic moment μ_1 given by

$$\mu_1 = \left(\frac{e}{T}\right)A \qquad ...(1)$$

where A is the area enclosed by the orbit.

Since the areal velocity of motion in a central orbit is $\frac{1}{2}r^2\left(\frac{d\theta}{dt}\right)$

$$A = \int_0^T \frac{1}{2}r^2\left(\frac{d\theta}{dt}\right)dt$$

Now the angular momentum pi is given by

$$p_l = mr^2\left(\frac{d\theta}{dt}\right), \text{ so that } \frac{1}{2}r^2\left(\frac{d\theta}{dt}\right) = p_l/2m$$

Hence
$$A = \int_0^T \left(\frac{p_l}{2m}\right)dt$$

Since p_l and m are constants

Substituting this value of A in equation (1)

$$\mu_1 = \frac{e}{T}\cdot\frac{p_l T}{2m} = \frac{e}{2m}p_l \qquad ...(2)$$

As p_l, according to the quantum theory, is equal to $l\left(\frac{h}{2\pi}\right)$,

$$\mu_l = \frac{eh}{4\pi m} l$$

Hence the orbital magnetic moment μ_l is directly proportional to the orbital quantum number l.

if $l = 0$, $\mu_l = 0$; if $l = 1$, $\mu_l = \frac{eh}{4\pi m}$, etc.

The quantity $\frac{eh}{4\pi m}$ represents the *smallest unit*, in terms of which the magnitude of atomic and sub-automic magnetic moments are measured. It is known as the *Bohr Magneton,* whose value can be readily computed from the known value of h and e/m; it is found to be 9.253×10^{-24} *SI Units.*

For various reasons, both theoretical and experimental, it has been found necessary to attribute *one Bohr magneton to the magnetic moment due to electron spin.*

The atom, with its magnetic moment arising from the orbital and spin motions of the electrons in it, may be regarded as an elementary magnet, whose dimensions, though small, are yet finite. If this atomic magnet be placed in a magnetic field, it will be acted upon by the field. But the nature of this action will depend on the nature of the field.

If the field is homogeneous, the atomic magnet will experience merely a couple which will rotate its axis into the direction of the field, since the external field acts upon its poles with equal and opposite forces. *If the atomic magnet moves in such a field* in a direction normal to the field, although its magnetic axis will be rotated into the direction of the field, it *will trace a straight line path* without any deviation, *i.e.*, without any transitional displacement.

On the other hand, *if the field is non-homogeneous*, the forces on the two poles will not be equal, which would result in a translatory displacement of the atom as a whole over and above the rotation of its axis into the direction of the field. *If then, the atomic magnet flies across such a non-homogeneous field* normal to the field direction, *it will be deviated away from its rectilinear path.*

An expression for the amount of the deviation thus produced may be obtained as follows:

Let the magnetic field be non-homogeneous along the X-direction, so that the field gradient is dB/dX and is positive. Let the atomic magnet, whose pole strength is p, length l and magnetic moment *M be placed* in the non-homogeneous field with its axis inclined at an angle 9 to the field direction.

If the field strength at one of the poles be B then at the other it will be

$$B+\left(\frac{dB}{dX}\right)l\ \cos\theta$$

Hence the force on the two poles of the atomic magnet due to the field are pB and p{B + (dB/dX). l cos θ} respectively. This means that, over and above the equal and opposite forces pH constituting the rotating couple, there is an extra force equal to p . (dB/dX). l cos θ acting on one of the poles alone; it is this force which displaces the atom as a whole; calling this translatory force F_x, we have

$$F_x = pl\left(\frac{dB}{dX}\right)\cos\theta = M\cos\theta\left(\frac{dB}{dX}\right) \qquad ...(3)$$

Supposing that the atomic magnet flies across the non-homogeneous field at right angles to the lines of force, it will be displaced from its straight path in the field direction. To find the amount of this displacement, let υ be the velocity of the atomic magnet of the mass m as it enters the field, L the length of its path in the field and t the time of flight through the field.

The acceleration α_x imparted to the atom along the field direction by the translatory force F_x is given by F_x/m.

The displacement D_x of the atom along the field direction at the end of time t is given by

$$D_x = \frac{1}{2}\alpha_x,\ t^2\frac{1}{2}\left(\frac{F_x}{m}\right).\left(\frac{L}{\upsilon}\right)^2$$

since t = (L/υ), the velocity of the atom perpendicular to the field being unaffected by the force in the field direction.

Substituting the value of F, from (1)

$$D_x = \frac{1}{2}.\frac{M\cos\theta}{m}.\frac{dB}{dX}.\left(\frac{L}{\upsilon}\right)^2$$

If μ be the resolved component of the magnetic moment in the field direction, then

$$D_x = \frac{1}{2} \cdot \frac{\mu}{m} \cdot \frac{dB}{dX} \cdot \left(\frac{L}{\upsilon}\right)^2 \qquad ...(4)$$

From this relation we see that if D_x, m, dB/dX, L and υ are known calculated. If dB/dX = 0, *i.e.*, if the field is made homogeneous, $D_x = 0$, *i.e.*, there, is no deviation.

Since $D_x \propto dB/dX$, the amount of deviation is determined by the degree of inhomogeneity of the field, in fact, to produce even a small measurable displacement; the inhomogeneity must be so marked that the field changes even within the small length (of the order of 10^{-8}cm.) of the atomic magnet. Stern and Gerlach succeeded in producing a sufficient inhomogeneity by a suitable construction of the pole-pieces of a magnet, one piece being shaped as a knife-edge, while the other had a flat face provided with a groove, as shown in the Fig. 1.31, the magnetic lines of force, in consequence, crowd together at the knife-edge, so that the field strength is considerably greater there that the other pole-piece.

Apparatus

The experimental arrangement devised by Stem and Gerlach diagrammatically in Fig. 1.31. A sampler of the substance to the examined is heated in an electric oven O and, n consequence, sends out atomic rays in all directions with a velocity corresponding to the temperature of vaporisation. By the use of slits S_1 and S_2 a sharp linear beam of atoms is obtained, which then passes between the specially shaped pole-pieces of the electromagnet M M.

The magnetic field is made as intense and non-homogeneous as possible, its lines of force being perpendicular to the direction of the beam. Finally the beam is made to strike a suitably prepared plate P, placed normal to the initial direction of the team. The entire apparatus is in an evacuated chamber. The arrangement, though simple in design, involves a very delicate technique in which a large number of details have to be carefully attended to.

The different parts of the apparatus are quite small. In some cases, the pole-piece is about 5cm long and the recording plate 3 mm square. The distance between the knife-edge and the plane of the slotted pole-piece is about 1 mm.

Experimental Procedure

In performing the experiment, the following precautions have to be taken:

(a) As the deflections obtained are small, the different parts of the apparatus must be carefully and accurately aligned.

(b) The oven must be capable of withstanding a high temperature and this must be under delicate control as regards constancy over long periods.

(c) The velocity of the atoms depends on the oven temperature which must be measured thermoelectrically or optically. For atoms in thermal equilibrium, the kinetic theory of gases gives the relation.

$$\frac{1}{2}m\upsilon^2 = \frac{3}{3}kT \text{ or } \upsilon = \sqrt{\frac{3kT}{m}}$$

where m and u are the mass and velocity of the atom, k the Boltzmann's constant and T the absolute temperature. Some preliminary experiments, however, showed that the mean velocity of the atoms issuing through slits was somewhat greater than this and hence the average value is to be taken as

$$\upsilon = \sqrt{\frac{3.5kT}{m}}$$

(d) In order to prevent collisions of the atoms in the beam with the atoms and molecules of any residual gas, the chamber must be exhausted for a long time, liquid air and charcoal being used.

(e) Accurate measurement of dB/dX is not easy. It is obtained by measuring the repulsion, at various points between the pole-pieces, of a thin bismuth wire mounted parallel to the edge of the knife-edge: pole-pieces. A correction is always found necessary for the variation in the value of dB/dX along the path of the deflected beam, as the beam is nearer the pole-piece at entrance into the field than at exist.

(f) Another problem is the development of the traces to render them Visible. Receiving plates of different kinds are used for the study of different atoms.

On account of these several technical difficulties to be overcome it is not surprising that of the many experiments begun, few were carried to a successful Theory of Atomic Structure.

Silver was the first to be investigated very completely, many series of experiments being made. With other elements, it was then possible to test whether the apparatus was functioning properly by interpolating experiments with silver.

Meissner and Scheffers improved the technique in the following manner which enabled the deviation D_x to be measured very accurately. Working with alkali metals, they caused the deflected beam to impinge on a heated tungsten filament instead of the receiving plate. The atoms of the alkali metals, on striking the hot filament, lost their electrons and were liberated as positive ions. The ionisation current thereby produced was measured in different positions and from the position of the maximum rent the magnetic deviation was accurately found.

Phipps and Taylor successfully carried out the difficult but important experimental determination of the magnetic moment of the hydrogen atom. The hydrogen atoms were produced in a long discharge tube under appropriate conditions or by dissociating the molecules by means of a hot filament. A stream containing a fairly high proportion of atomic hydrogen was passed through three special slits in a glass tube between which two high vacuum pumps were operating so that tine emergent beam contained effectively free, atoms. This atomic beam after passing through a powerful non-uniform magnetic field was received on a plate coated with molybdenum trioxide. The atoms reduce the oxide and so produce a visible trace.

A sharply defined blue line against a white background was the result in the absence of the filed, while when the field was on, the beam was separated into two, and two traces were obtained. For determining the exact positions of the deviated lines microphotographs were taken. From the deviation thus, measured the magnetic moment of the hydrogen atom was calculated to be one Bohr magneton within the limits of experimental error.

Results

In the case of silver, with the field off, a fairly sharp line was traced by the atomic beam on the receiving plate. On establishing the non-homogeneous field a double trace was obtained, one on either side of the original trace. The traces obtained were somewhat as shown in Fig. 1.32. The irregularity in the right hand trace is due to the irregularity of the magnetic field near the knife-edge pole-piece. The convergence at the top and bottom arises from the fact that the field gradient decreases transversely. The traces actually obtained are much smaller than what is shown here, but they can be magnified for easy measurement. A certain diffusiveness is unavoidable owing to the velocity distribution (Maxwellian) of the atoms, but the magnetic splitting of the beam is relatively great enough, that the diffusiveness becomes of secondary importance. The mean deviation D_x of the beam caused by the field can be measured from the traces with fair accuracy. Thus, D_x being known, and the other quantities such as dB/dX, L and u being readily measured, μ can be calculated. In the case of silver, it was found to be equal to one Bohr Magneton. The results, obtained with other elements experimented upon, may be summarised as follows:

Cu, Au, H, Li, Na, K: all gave double traces, the separation leading to the value of $\mu = \pm 1$ Bohr magneton, like silver.

Zn, Cd, Hg, Sn, Pb: no effect was produced by the field. Hence $m = 0$.

Tl (Thallium): the trace was divided into two, there being a definite absence of undeviated atoms. The deviation gave the value of

$$\mu = +1\frac{1}{3} \text{ Bohr magneton.}$$

Sb, Bi: Sb showed only an undeviated trace. The Bi trace obtained by Stern and Gerlach was remarkable in showing a one-sided continuous broadening indicating not only undeviated but also strongly attracted carriers. But later, a symmetrical splitting was obtained by Leu who showed that the distribution of density was consistent with the presence of atoms having values for μ in the ratio of 1 : 3.

Ni, Co, Fe: Nickel showed curious results in that three clearly defined traces appeared, one of them in the undeviated position itself, which meant that some of the atoms have no magnetic moment ($\mu = 0$). For the deviated traces μ was somewhat greater than one magneton. In addition to the three pronounced traces, there were indications of the presence of the atoms might having a value of μ greater than six. In the case of cobalt, a value $\mu = 6$ could just be established.

INTERPRETATION OF RESULTS

It is important to consider, in the first place, whether the beam consists of single atoms or molecules, since the experiments directly give the resolved moments of the carriers, whether atoms or molecules. At the temperature employed, there is little doubt that the vapours of Cu, Ag, Au, Sn, Pb, and Tl are monoatomic, so that the values of μ deduced for them do correspond to atomic moments. Ni, Co, Fe also probably monoatomic. But with Bi and Sb there is a large proportion of molecules in the beam.

Considering only the sure cases, it is possible to show how the results obtained offer a beautiful confirmation of the fundamental postulates of the vector model:

(a) Spatial Quantisation

The classical theory lays no restriction, on the orientation of the atomic magnet with respect to the field direction, all values from 0° to 180° for the angle 6 in the relation $\mu = M \cos \theta$ being permissible. Hence the displacement of the atoms in the non-homogeneous field should cover a continuous range, so that a *continuous diffuse band* should he obtained on the receiving plate. Further, on account of the Maxwellian distribution of velocities of the atoms in the beam, the densest part of the trace should be in the centre, since the number of atoms whose axes make angles between θ and $(\theta + d\theta)$ with the field is nearly proportional to $\sin \theta$ and hence a large fraction of the atoms would be oriented nearly transversely to the field and so would suffer very small deflections.

On the other hand, *according to the spatial quantisation theory*, not all settings are possible for the atomic magnet with respect to, the field direction, but only a certain discrete number. If the total angular momentum number of the atom is J the permitted orientations are $(2J + 1)$. Hence on the plate, instead of obtaining a continuous band, $(2J + 1)$, distinct traces should be obtained. Further, if the atom under test belongs to the *one electron, system* and is in the *ground state*, one can have higher values of J, since although $L = 0$, $S = 1/2 = J$, so that $(2J + 1) = 2$. In such a case, therefore, only two orientations are permitted and hence two traces should be obtained on the plate. In the case of many *electron system*, the atoms in the *ground state* can have higher values of J, since although $L = 0$, S can assume values 0, 1/2, 3/2, 2, etc., and $J = S$. The number of permitted orientations may be greater than two, so that more complex setting of the beam could be expected: *e.g.*, for atoms of $^6S_{5/2}$ term J being equal to 5/2, six traces should be obtained on the plate.

All these Theory of Atomic Structures of the spatial quantisation theory are exactly verified by the results of the Stern and Gerlach experiment. Thus,

(i) *H, Li, Na, K, Cu, Ag, Au, all belonging to the first column of the periodic table*, have In the ground state the spectral term $^2S_{1/2}$, which means that 2 J + 1 = 2. They all give only two traces as expected from theory. As soon as the atoms of these elements enter the magnetic field their magnetic axes orient themselves in the two permitted directions parallel and antiparallel to the field direction, no intermediate position being allowed. Further, in this case it can be shown that $\mu = \pm 1$ Bohr magneton, actually obtained from experimental data.

(ii) *Zn, Cd, Hg, belonging to the second column*, have two s electrons outside closed shells, and their normal state, as indicated by their spectra, is $^1S_0$. This means that J = 0 and hence the atom as whole has no resultant angular momentum and no magnetic moment. That is why no effect is produced on the atoms of these elements by the field.

(iii) Tl belonging to the third column, has one p electron in addition to completed groups. This gives a $^2P_{1/2}$ ground term *i.e.*, L = 1, J = 1/2, 2J + 1 = 2. Hence it gives rise to two traces and calculations show that m = 0, which agrees with the experimental results.

(iv) Sn, Pb, belonging to the *fourth column* have two p electrons not forming a completed group, but the total magnetic moment vanishes in their normal $^3P_0$ state. This is why the atoms of these elements are undeviated by the field in the experiment.

(b) Electron Spin

Considering the well established case of silver, experiment clearly shows that the atomic beam of silver is split up by the magnetic field into two parts of approximately equal intensity. Further the value of the magnetic moment obtained is ±1 Bohr Magneton. From theory it is known that for the normal state of the silver atom L = O. Hence, if the spin of the electron does not exist, S would be zero, so that J also would be equal to zero and there would be no splitting of the atomic beam into two components. If, on the contrary, the existence of electron spin is admitted and the value of 1/2 be given to S in this case, there is perfect agreement between theory and experiment.

(c) Atomic and Quantum Nature of Magnetism

The idea of atomic nature of magnetism goes back to the days of Weber. The view that paramagnetic substances have a permanent molecular magnetic moment, while diamagnetic substances possess no such moment is a long established fact in the physical theory of magnetism. Weber was the first to develop this idea on the molecular current hypothesis of Ampere. It was next rendered certain by Langevin's treatment of paramagnetism based on the kinetic theory of gases and of diamagnetism on the basis of the classical electron theory. Weiss, in 1911, from experimental data then available, concluded that there was a fundamental unit of magnetic moment, of which all atomic and molecular moments were multiples. This unit was called after him the *Weiss Magneton,* whose value was estimated to be 1123.5 gauss cm. per gram-atom or per gram-molecule, which gives 1.85×10^{-21} gauss-cm. per atom or per molecule. With the advent of the quantum theory, another fundamental unit, known as the *Bohr Magneton,* came to be recognised. Its value was found to be 5,590 gauss-cm. per gram-atom or 9.21×10^{-21} gauss-cm. per atom, which is therefore, almost five times as large as the older unit.

The Stern and Gerlach experiment not only confirms the general Theory of Atomic Structures of the classical theory about para and diamagnetism, but goes much further and establishes the quantum nature and atomic origin of magnetism. For, in the first place, according to the results of the experiments, substances which are found to be diamagnetic have no atomic magnetic moment ($\mu = 0$), *e.g.*, Zn, Cd, Hg. Pb, Sn, etc. Substances which are paramagnetic are made up of atoms with one valency electron in the ground state, thereby giving rise to a magnetic moment of 1 Bohr Magneton due to the spin of the single free electron: *e.g.*, H, Li, the alkalis, Cu, Ag, Au. Ferromagnetic substances (Fe, Ni, Co) are made up of atoms with intermediate incomplete electronic shells and have a large value for the atomic moment.

Secondly, the results establish also that the true fundamental unit is the Bohr magneton and not the Weiss magneton. The latter based on the classical theory, according to which all possible orientations of the atomic axis with respect to the field are permissible, has only an apparent existence. The former is the one given by the experiment of Stern and Gerlach and conforms to the quantum theory; it alone, therefore, has a real existence.

There is no doubt that the vector atom model is a superb and many sided conception, which enables one to solve most of the important and intricate problems concerning atomic structure and spectral phenomena.

WAVE MECHANICALATOM MODEL

The great drawback of the vector atom model is that it does not present itself as a single unified system, containing the necessary theoretical justification of its principles and postulates. These are introduced somewhat piecemeal, partly on empirical data and partly by analogy. No adequate reasons are given for the use of electron spin with half-integer value; the assumption of discrete energy states and of emission of radiation when the atom passes from one of these states to another are arbitrary as in older models. The model therefore lacks a sound theoretical foundation. The most recent theory, known as the *wave mechanical atom model*, supplies this information and presents a picture of the peripheral electronic structure of the atom, the best that can be had in the present state of our knowledge.

A detailed study of this model is beyond the scope of this book. A masterly treatment by *Dirac* of the relativistic theory of the electron based on wave mechanics offers, almost without any special effort, theoretical justification of the postulates and Theory of Atomic Structures of the vector model. We shall limit ourselves to a brief survey of the essential features of this new conception, indicating the advances made over the older models.

According to wave mechanics, in order to have a complete picture of the electron, we should consider it under the double aspect of *particle* and *wave*. The change of finding a moving electron at a given point is therefore governed by a wave, which necessarily involves a certain degree of uncertainty about this exact position. The older models, which placed the electron at a precise point is a well-defined orbit, did not take into account this correct nature of the electron and to this fact all their short-comings are ultimately to be traced.

The-wave mechanical theory of the atom can be explained in a simple manner, *i.e.*, without entering into complicated mathematical treatment, analogy with other forms of waves. Free electrons emitted from some source and spreading outwards through vacuum are like the ripples on the surface of a large pond, which advance from the source of disturbance in ever-widening circles, becoming feebler as the distance from the starting point increases, until finally they are hardly perceptible. On the other hand, electrons in an atom, experiencing the electrostatic attractive force of the nucleus and hence prevented from leaving its neighbourhood, are like the stationary wave set up by the reflection of the ripples at the walls of a narrow vessel in which the ripples are produced, or better still like the stationary waves set up in a fixed string with a number of loops and nodes giving the various modes of vibration. On this analogy the electron waves may be said to be reflected back when they reach the boundary of the atom so that a system of stationary waves may be imagined within the atom.

With this picture of the atom a number of important Theory of Atomic Structures may be drawn:

1. Removal of Individualistic Particle Picture of the Electrons in the Atom

Giving to the electron a fixed position in a definite orbit is no longer correct. The electron probability is distributed over the whole range of the atom though corresponding to die loops and nodes in the stationary wave system the distribution is not uniform. For the same reason, when several electrons are contained in the atom, they lose their individualities to a certain extent, as it is not possible to say that any particular part of any of the "probability waves" representing the electrons belongs to any particular electron.

2. Nebulous Picture of the Atom

The atom loses much of its former vividness of well-defined electronic orbits, since the electron probability does not vanish sharply at any distance from the nucleus; it becomes zero only at infinity, so that theoretically there is a chance of finding the electron anywhere in space. But there is a region where the probability falls steeply to a small value and this corresponds to what is ordinarily called the boundary of the atom.

3. Quantum Stationary States of the Atom

The stationary electron waves can vibrate in different modes which imply different wavelengths and frequencies, and therefore, different amounts of energy in the system. In the mathematical treatment of the problem, this means that the wave equation representing the stationary .waves in the atom can be solved only when certain coefficients are given appropriate values, which must be whole numbers or occasionally simple fractions, like the different harmonic of a fundamental vibration. These coefficients: correspond to the *quantum numbers* of the older models, but they appear here quite naturally as characteristic solutions of the wave equation, whereas in the older theories they were introduced somewhat artificially, with the only reason that they were found necessary for a satisfactory interpretation of the experimentally observed atomic and spectral phenomena.

4. Non-Radiative Character of the Stationary States

In a stationary wave system the amplitude at any given point is constant and does not change with time. Now, considering the wave mechanical atom from the point of view of what is known as the *electronic cloud round* the nucleus or the density distribution of the electric charge in the atom, and interpreting square of the amplitude of the electron wave as measuring the electric charge density, it follows that the charge density at any point must be no matter how great the proper energy value, the charge distribution must be static, *i.e.*, must not vary with time. There being no movement of electric charge in the permitted energy states, the atom in such states cannot radiate electromagnetic energy, in consonance with classical electrodynamics. Thus, we arrive quite naturally at a result, which in the older theories had to be assumed as an arbitrary postulate contrary to the classical theory.

5. Emission of Radiation Governed by Bohr's Frequency Condition

When two vibrations of somewhat different frequencies are superimposed they give rise to the so-called "beat" phenomena, in which the amplitude waxes and wanes with a frequency equal to the difference between the two superposed frequencies. In a similar manner, when two different modes of the wave system, corresponding to two permitted energy states of the atom, can be excited simultaneously so that they are superimposed, radiation whose frequency is given by the number of beats per second between the frequencies of the two electron waves superposed, will be emitted. For, if we once again interpret the square of the amplitude of vibration in the wave mechanical model as a measure of the electric charge density, it is readily seen that when different energy states are simultaneously excited, the charge density at any given point is no longer constant but varies with time.

Hence, there is a variation of the electric charge, which must give rise to the emission of electromagnetic waves. The frequency of these emitted waves will be the best frequency. If v_L be the frequency of radiation emitted and v_1 and v_2 the frequencies of the two superposed vibrations of electron waves, then:

$$v_L = v_1 - v_2$$

According to the wave mechanical theory the energy W of a stationary state is given by W = hv, where v is the frequency of the electron waves corresponding to that state. Hence $v_1 = W_1/h$ and $v_2 = W_2/h$ where W_1, and W_2 are the energies of the two states involved.

$$v_L = \frac{W_1 - W_2}{h}$$

Thus, we arrive logically at Bohr's frequency condition which was introduced arbitrarily as a fundamental postulate of his theory, at variance with the classical standpoint.

Thus, the wave mechanical atom model is able to account adequately for all the postulates assumed by the older models. Further, it has the supreme merit, in a scientific theory of reducing the number of fundamental postulates and presenting itself as a single unified system.

SOME EXAMPLES OF ELECTRON CONFIGURATIONS

In order to illustrate the theory proposed above, the electronic configurations in some of the elements will now be considered.

Hydrogen (Z = 1) has only one electron which will naturally occupy the first K shell for which n = 1, in order to give least energy to the system. According to Pauli's Principle, there can be two electrons in the **K** shell, or the single electron can exist in two states with $m_s = +1/2$ and $m_s = -1/2$. On account of the single electron which cannot complete the **K** shell, one may expect the atomic hydrogen to be very active chemically, combining its single electron with another hydrogen atom to form a stable helium-like pair, the molecule of hydrogen, which is known to be comparatively inactive.

The symbolic representation of the electron configuration in the hydrogen atom is 1_S, 1 referring to the total quantum number of the electron and s the type of electron, since $l = 0$ in this case. This is confirmed by the fact that hydrogen is *monovalent and electropositive.*

Helium (Z = 2) has two electrons, both of which, according to Pauli's Principle, can occupy the first **K** Shell. Both spectroscopic and ionisation potential data show that to be really the case. Further, the two electrons complete the shell, which renders the helium atom an *inert* monotonic gas, completing the first period of the periodic table of elements. The normal helium atom is therefore represented by $1s^2$, *i.e.,* two electrons for which n = 1 and $l = 0$, the total number being affixed at the top of s and the rectangular enclosure indicating that the electrons are interlocked in a closed shell. The spark spectrum of helium resembles the arc spectrum of hydrogen. This is what we should exact from theory as the singly ionised. He atom found in the spark spectrum has a hydrogen-like structure, consisting of a positive nucleus about which a single electron rotates; but the frequencies of the corresponding lines in the helium spectrum are higher than in hydrogen, because the stronger nuclear charge in He causes all energy levels lie much lower.

Lithium (Z = 3) has three electrons, two of which occupy the **K** shell and the third the next **L** shell, in order to keep the energy of the system at a minimum. The third electron in the incomplete **L** shell is the valency electron. Hence the neutral normal atom of Li is represented by $1s^2$ 2s. *Optical* spectrum of Li shows that the lowest orbit is such that this third electron is really 1s and not 2p or any other. This valency 2s electron is much more free, *i.e.,* much more loosely bound than the $1s^2$ electrons, as is seen from the fact that Li is strongly *monovalent and electro-positive. The arc* spectrum of Li is predominantly that to be expected from a single electron system of the alkali type. The spark spectrum, on the other hand, which is ascribed to singly ionised Li is similar to the arc spectrum of He, for the same reason as that given above for the singly ionised helium.

Sodium (Z = 11) has a total number of 11 electrons. According to Pauli's principle, 10 of them will be interlocked in the first two **K** and **L** shells, completed by 2 and 8 electrons respectively, so that the remaining single electron will have to occupy the third **M** shell. The two electrons in the **K** shell are represented by $1s^2$. Of the eight electrons in the **L** shell, will be in the first sub-group ($l = 0$) and hence represented by $2s^2$ the remaining six in the second sub-group ($l = 1$) and hence represented by 2p. The last electron in the **M** shell is the valency electron which is found to be as electron from energy considerations and spectroscopic data; it is therefore represented by 3s. The electron configuration of the normal sodium atom is represented by

$$1s^2\ 2s^2\ 2p^6\quad 3s$$

The arrangement receives confirmation from the fact that sodium is monovalent and electropositive,

Magnesium (Z = 12) has 12 electrons of which 10 occupy the first two shells, while the remaining two will have to go to the third M shell. These are the valency electrons, making Mg divalent. A study of the arc and spark spectra show that the two valency electrons, are,} electrons. Hence the normal Mg atom is represented, by

$$1s^2\ 2s^2\ 2p^6\quad 3s^2$$

Aluminium (Z = 13). Leaving aside the 10 electrons interlocked in the first two shells, the remaining 3 occupy the third M shell. Hence, Al is trivalent. Spectroscopic study and energy consideration show that the eleventh and twelfth electrons are of the s type, while the thirteenth is a p electron. The symbolic representation of Al atom is

$$1s^2\ 2s^2\ 2p^6\ 3s^2\quad 3p$$

Copper (Z = 29). Of the 29 electrons, 2 will occupy the first K shell, 8 the two sub groups of the second L shell and 18 the third M shell, distributed as 2, 6 and 10 in the three sub-groups. What is left over is only one electron, the valency electron which occupies the fourth N shell. Hence copper should be monovalent and its electronic configuration is

$$1s^2\ 2s^2\ 2p^6\ 3s^2\ 3p^6\ 3d^{10}\quad 4s$$

The arc spectrum of copper should therefore resemble that of the alkalis like Li, Na, etc. It is to be noted however, that the last of the electrons, which completes the 3d sub-shell is rather lightly bound and comes off easily so that copper frequently a valency of two as an alternative to the expected valency of one, as readily seen from the existence of both cupric oxide (CuO) and cupric oxide (Cu_2O).

Thallium (Z = 81). The total number of electrons is therefore 81. The normal atom of Tl is represented applying Pauli's Principle and paying due attention to energy considerations, by

$1s^2$	$2s^2 2p^6$	$2s^2 3p^6 d^{10}$	$4s^2 4p^6 4d^{10} 4f^{14}$	$5s^2 5p^6 sd^{10}$	$6s^2 6p$
K	L	M	M	O	P

Hence it has the first four shells complete. But before the fifth O shell is completed, the sixth P shell is started. Of the three electrons in the last shell, two are s electrons and one p electron. Hence Tl is *trivalent.* These few examples will serve to demonstrate the methods and some of the fundamental principles employed in ascertaining the distribution of electrons in atoms in the different elements. The same procedure is followed in the case of any element in the periodic table.

FINE STRUCTURE OF SPECTRAL LINES

The Selection Rules

It is experimentally found that all the possible combinations of permitted energy states of an atom do not actually appear as spectral lines. Selection rules are certain principles, which give the reason for such a state of affairs.

The first selection rule devised in connection with the older atom model referred to the azimuthal quantum number n ϕ or k and it was $\Delta k = \pm 1$ as we have seen. This was formulated on an empirical basis, but was later justified theoretically by Rubinowicz by applying the theorem of an emission of electromagnetic radiation by an atom. This simple rule, however, was soon found inadequate.

For the vector atom model, three selection rules have been devised, one for L, another for J and third for S.

(a) *The selection rule for L*: Most of the observed spectral lines are due to transitions between states, in which a single electron jumps from one orbit to another, and in such cases, the selection rule is $\Delta L = \pm 1$, *i.e.*, only those lines are observed for which the value of L changes by ± 1.

(b) *The selection rule for J*: Spectral lines arise only when transitions lake place between states for which $\Delta J = \pm 1$ or 0, $0 \rightarrow 0$ being, however, excluded.

(c) *The selection rule for S* is given by $\Delta S = 0$, which means that states with different S (hence different multiplicities) do not combine with one another. Theory and experiment, however, show that this selection rule is adhered to less and less strictly as the atomic number increases. Hence, it is only an approximate rule holding good in the case of light atoms.

Note: In the presence of a magnetic field, the orbital magnetic quantum number, m_L either does not change or changes by ±1 *i.e.*, $m_L = 0$ or ±1. The spin magnetic quantum m, remains unchanged, *i.e.*, $\Delta m_s = 0$. Inconsequence

$$\Delta m_J = 0 \text{ or } \pm 1$$

These selection rules were first introduced on a purely empirical basis in the study of optical and X-ray spectra and Zeeman Effect. Later, they were obtained by a rigorously deductive method in wave mechanics and hence they now rest on a permanent theoretical basis.

These rules furnish invaluable help in the allocation of observed spectral series to the proper quantum number. With their aid, energy level diagrams can be constructed both for the natural complex multiple lines and for the Zeeman Effect of such lines.

Transitions which contravene these rules are sometimes observed, but the intensities of such "*forbidden lines*", as they are called are usually weak in comparison with the normal lines.

The intensity rules have been devised to supplement the selection rules, in order to predict also the intensity of the lines that occur. These were originally postulated on an empirical basis in the study of optical and X-ray spectra. Later, a full theoretical derivation was given on a wave mechanical basis. The intensity rules are :

(a) *Those transitions are strong, giving rise to intense lines, in which L and J change in the same sense: the transitions are the weaker the more the change in direction of L and J is different.*

(b) *A transition in the decreasing sense ($L \rightarrow L - 1$) is, ceteris paribus, stronger than a transition in the increasing sense ($L \rightarrow L + 1$).*

(c) The case of oppositely directed transitions does not occur, in general, either in X-Ray spectra or in subject spectra; because it would lead to a final state, in which (J – L) would be two units greater than in the initial state, which is forbidden.

Hence we may distinguish the following cases:

$$\Delta L = \pm 1, \Delta J = -1\text{: most intense line (a)}$$

$$\Delta L = -1, \Delta J = 0\text{: less intense (a)}$$

$$\Delta L = +1, \Delta J = +1\text{: weaker (b)}$$

$$\Delta L = +1, \Delta J = 0\text{: weakest (a) and (b)}$$

$$\Delta L = -1, \Delta J = +1$$

$$\Delta L = +1, \Delta J = -1$$

Note: It can be shown both theoretically and experimentally that the total intensity of all the lines coming to or starting from any given J term of a multiple is proportional to (2J + 1).

The Interval Rule

Lande discovered a rule regarding the interval in frequency between the different levels constituting a multiple. It is called the *Lande interval rule* and states that the frequency interval between two levels with total angular momenta (J + l) and J respectively is proportional to (J + l).

There are, however, many deviations from this interval rule, chiefly in cases when the coupling scheme follows neither the L-S type, nor the j – j type, but is intermediate between these extreme cases.

Spectral Terms

Since a spectral line arises due to the transition of the atom from one energy state to another, in the interpretation of the fine structure of a spectral line, one must necessarily start with the determination of the different possible energy states. Then by the use of adequate selection rules the states between which transitions actually take place are picked out. Such transitions should account for the observed fine structure.

The different energy states of the atom between which transitions can take place have come to be known by the special name in spectroscopic study, *viz., spectral terms*. The problem of the fine structure is thus essentially a problem of spectral terms. When the latter is solved, the former is readily understood. Hence in the application of the vector atom model to the fine structure phenomenon, it must, first of all be seen how the model can be used to fix up the spectra terms involved. For the sake of clearness, the fine structure in optical and X-ray spectra will be considered separately, as the notations used in the two cases are different.

OPTICAL SPECTRA

Spectral Terms

In the classification of the spectral terms by the use of the vector atom model, it is convenient to divide atoms into two main categories, *viz. one-electron system and many-electron system.*

The atoms belonging to the first class have *only one valence or optical electron, e.g.*, hydrogen and hydrogen-like atoms, such as the alkalis. Even if the atom of this system actually contains several electrons, all of these, except one, are interlocked in closed shells, which, in consequence, do not contribute to the total angular momentum of the atom, according to Pauli's Principle. Hence, the single free electron in the outermost shell is alone effective in determining the spectral properties of the atom.

The atoms of the second class have *more than one valence or optical electron, i.e.*, several free electrons not interlocked in closed shells. They will, therefore, contribute their share to the total angular momentum of the atom and hence become effective in fixing the spectral properties, *e.g.*, the alkaline earths, belonging to the two electron system, aluminium, scandium, etc., belonging to the three electron system, titanium to the four electron system and so on.

Considering the one-electron system, since the state of the atom a whole is determined by that of a single free electron, the values of l, s and j referring to the electron are equal to those of L, S, J, defining the whole atom. Hence $S = s = \pm 1/2$ and the multiplicity of state given by $r = (2S + 1)$ is two. This system *will therefore give rise only to doublet terms with the exception of the ground terms* which corresponds to the normal state of the atom, *i.e.*, to $L = 0$. For, in present case, the possible values of J are only two *viz.*, $J = (L + 1/2)$ (and $L - 1/2$). Now, if $L = 0$, $J = \pm 1/2$; but since the net angular momentum of the atom is always a positive quantity, the only value of J in this case is $J = + 1/2$, which means a *singlet*. If $L = 1$, $J = 1 \pm 1/2$. *i.e.*, 3/2, or 1/2 (*doublet*). For $L = 2$, $J = 2 \pm 1/2$, *i.e.*, 5/2 or 3/2 (*doublet again*) and so on.

In the *many electron system,* on the other hand, the state of the atom will be determined by the states of several free electrons. In consequence, the vectors L, S and J of these electrons. Hence the value of S is not always 1/2, but may be 0, 1/2, 1, 3/2, 2, etc., according to the number of free electrons involved and their orientations, parallel or anti-parallel, and the multiplicity r is no longer two in all cases, but is 1 when S = 0, 2, when S = 1/2, 3 when S = 1, 4 when S = 3/2, 5 when S = 2 and so on. Thus, the possible values of J for a given value of L may be one, two, three, four, etc., which means *each state may be a singlet or multiple.*

In the *two-electron system,* since S = 0 or 1, the state can be only a *singlet* or *triplet.* In the *three-electron system,* as S = 1/2 or 3/2, each state can be & doublet or a quartet. In the *four-electron system* since S = 0, l or 2, *singlet, triplet and quintent* are possible. In general, *odd electron systems* have even number of terms, while *even electron systems have odd number of terms.* Accordingly, the number of spectral terms of successive atoms in the periodic table of elements will alternate between even and odd.

But the ground terms is always a singlet in the system also. For, it corresponds to L = 0 and J = S. Now, since S can assume values 0, 1/2, 1, 3/2; etc., S is never less than L. For S = L = 0, the possible value of J is one ; when S > , the multiplicity of the state is given by (2L + 1) which leads again to the possible value of J as one.

Limitation of Term Multiplicity due to Pauli's Principle

Pauli's Exclusion Principle causes, however, the multiplicity of terms not to attain always the maximum (2S + 1) values which are to be expected from the number of free electrons involved. Taking, for instance, the case of an atom with 6 electrons in the L shell, at first sight, one expects the multiplicity r = (2S + 1) to attain the maximum value of 7 on the assumption that the spins of all the six electrons are similarly directed there by leading to a value of 3 for S. But such an arrangement of spins cannot take place without violation of Pauli's Principle. For, in this case, n = 1, l = 0 and 1; m_l can have only four values, *viz.*, 0 corresponding to l = 0, and 1, 0, –1, corresponding to l – 1. Now if we associate with these 4 values of m_l *six like* values of m_l *i.e.*, all either +1/2 or –1/2, the result would be that more than one electron will have identical quantum numbers as shown below :

Hence only four of the six electrons can have similarly directed spins (*i.e.*, same sign for m_s), while the other two must be directed oppositely. In consequence, the value of S reduces to 1, from which it follows that the actual multiplicity is only 3. Another inference is that, as a shell is gradually filled by the addition of successive electrons, the term multiplicity first rises and then falls, the variation being symmetrical with respect to the middle of the shell.

L_I (n = 2, l = 0)	L_{II} (n = 2, l = 1)
$m_l = 0$	+1, 0, –1
m_s = +1/2, +1/2	+1/2, +1/2 +1/2 +1/2
The correct arrangement is	
$m_l = 0$	+1, 0, –1
m_s = +1/2, –1/2	+1/2, +1/2, +1/2, –1/2

Notations of Spectral Terms

The states of the atom, in which the values of its L vector are 0, 1, 2, 3, 4, etc., are symbolically represented by the capital letters, S, P, D, F, G, etc., respectively. The value of J is appended to this symbol as a subscript and the multiplicity of the state r written as a left superscript.

Thus, the spectral terms corresponding to L =1 and S = 1/2 are written as $^2P_{1/2}$ and $^2P_{3/2}$ For, since L = 1, the capital letter which represents the state is P, the multiplicity of the state r = (2 × 1/2 + 1) = 2 which is put as a superscript; since J = L ± S = 1 ± 1/2, *i.e.*, 1/2 or 3/2, these are put as subscripts.

To take another examples, the notation $^4D_{5/2}$ specifies a spectral term belonging to a state whose L = 2, J = 5/2 multiplicity r = 4 and hence S = 3/2 (from r = 2S + 1). Since the possible values of J are four, *viz.*, 7/2, 5/2, 3/2, 1/2, the above notation refers only to one of the four terms of the given state.

In the atoms belonging to the one-electron system, since every state is a doublet, except the ground state, the notations for the spectral terms in these atoms are

$$^2S_{1/2},\ ^2P_{1/2},\ ^2P_{3/2},\ ^2D_{3/2},\ ^2D_{5/2},\ ^2F_{5/2},\ ^2F_{7/2},\ \text{etc.}$$

In the many electron system atoms, the notations will be more complex, but they can be readily written, provided the necessary data are available.

It may be noted that the *multiplicity symbol of the system is used on all terms even if all of them are not present.* Thus, the ground term of all alkali atoms is written as $^2S_{1/2}$, although the multiplicity is one and we should strictly speaking write $^1S_{1/2}$, . But this method has the advantage of indicating to which system the ground term belongs; in the present case, for instance, to the one-electron system, giving rise to doublet terms in general.

It is likewise important to note that the vector model, where only the vectors L, S and J are involved, cannot determine the magnitude of the energies corresponding to the different spectral terms, since, the total quantum number n, necessary for such a determination, does not directly enter in the fixing up of the spectral terms. Yet the three vectors, L, S and J characterise the different energy states in such a definite manner that they may be used, without the intervention of n, to predict which transitions between the various energy states are permitted ;and hence which spectral lines may be expected.

SOLVED EXAMPLES

Example 1:

The wavelength of the first line of Balmer series is 563 Å. Calculate the wavelength of the second line of Balmer series.

Solution:

For the first member,

$$\bar{v}_1 = R\left[\frac{1}{2^2} - \frac{1}{3^2}\right]$$

$$\bar{v}_1 = \frac{5}{36}R$$

or
$$\frac{1}{\lambda_1} = \frac{5}{36}R \qquad \text{...(i)}$$

For second member

$$\bar{v}_2 = R\left[\frac{1}{2^2} - \frac{1}{4^2}\right]$$

$$\bar{v}_2 = \frac{3}{16}R$$

or
$$\frac{1}{\lambda_2} = \frac{3}{16}R \qquad \text{...(ii)}$$

Dividing (i) by (ii)

$$\frac{\lambda_2}{\lambda_1} = \frac{20}{27}$$

or $$\lambda_2 = \left(\frac{20}{27}\right)\lambda_1$$

But $\lambda_1 = 6563 \times 10^{-10}$m

$$\therefore \quad \lambda_2 = \left(\frac{20}{27}\right) 6563 \times 10^{-10}\ \text{m}$$

$$\lambda_2 = 4861 \times 10^{-10}\text{m}$$

$$\lambda_2 = 4861\ \text{Å}.$$

Example 2(a):

The radiations given off by the Hg atoms returning to their normal states were studied by Frank and Hertz and a line was observed at 2537Å. Calculate excitation potential for Hg.

Solution:

Here $\lambda = 2537\ \text{Å} = 2537 \times 10^{-10}\text{m}$

Let excitation potential be V

$$\therefore \quad eV = h\nu = \frac{hc}{\lambda}$$

$$\therefore \quad V = \frac{hc}{e\lambda}$$

Here $h = 6.624 \times 10^{-34}$ J-s

$c = 3 \times 10^8$ m/s

$e = 1.6 \times 10^{-19}$ C

$\lambda = 2537 \times 10^{-10}$m

$$\therefore \quad V = \frac{6.624 \times 10^{-34} \times 3 \times 10^8}{1.6 \times 10^{-19} \times 2537 \times 10^{-10}}$$

$$V = 4.9\ \text{volts}.$$

Example 2(b):

What are the energy, momentum and wavelength of the photon emitted by hydrogen atom in transition 2s → Is?

Solution:

In hydrogen,

Energy of 1s state, $E_1 = -13.6$ eV

Energy of 2s state, $E_2 = \frac{-13.6}{2^2} = -3.4$ eV

(1) Energy of photon,

$$E = E_2 - E_1 = -3.4 - (-13.6)$$

$$E = 10.2 \text{ eV}$$

$$E = 10.2 \times 1.6 \times 10^{-19} \text{ J}$$

$$E = 1.632 \times 10^{-18} \text{ J.}$$

(2) Momentum $= \frac{E}{c} = \frac{1.632 \times 10^{-18}}{3 \times 10^8}$

$$= 5.44 \times 10^{-27} \text{ kg–m/s.}$$

(3) $E = hv = \frac{hc}{\lambda}$

$$\lambda = \frac{hc}{E}$$

$$\lambda = \frac{6.624 \times 10^{-34} \times 3 \times 10^8}{1.632 \times 10^{-18}}$$

$$\lambda = 12.18 \times 10^{-8} \text{ m} = 1218 \times 10^{-10}\text{m}$$

$$\lambda = 1218 \text{ Å.}$$

Example 2(c):

Calculate the energy required to create a vacancy in

(i) K shell of copper atom

(ii) L shell of copper atom

Given, Ionization potential of hydrogen atom = 13.6 eV

For copper Z = 29.

Solution:

(i) The energy required to create a vacancy in the first orbit (n = 1) *i.e.,* K shell of copper atom.

$$E_1 = \frac{(Z^2) \times 13.6}{n^2}$$

$$= \frac{(29)^2 \times 13.6}{1}$$

$$= 11.44 \times 10^3 \text{ eV}.$$

(ii) Energy required to create a vacancy in the second orbit (n = 2) *i.e.,* L shell of copper atom

$$E_2 = \frac{(29)^2 \times 13.6}{2^2}$$

$$E_2 = 2.86 \times 10^3 \text{ eV}.$$

Example 2(d):

In a μ - mesonic atom, a muon with a charge-e and mass 200 m_e moves in a circular orbit around the nucleus of charge + 3e. Assuming that Bohr's model of atom is applicable to this system.

(i) Derive the expression for the radius of n th Bohr orbit

(ii) Find the value of n for which the radius of the orbits is approximately the same as that of Bohr's first orbit of hydrogen atom. Bohr radius = 0.53×10^{-10}m.

Solution:

(i) Here $Z = 3$

$m = 200m_e$

Here $$\frac{m\upsilon^2}{r} = \frac{Ze^2}{4\pi \epsilon_0 r^2}$$

$$m\upsilon^2 = \frac{Ze^2}{4\pi \epsilon_0 r} \qquad ...(i)$$

Also $$m\upsilon r = \left(\frac{nh}{2\pi}\right) \quad \text{...(ii)}$$

$$\therefore \quad \upsilon = \left(\frac{nh}{2\pi\, mr}\right)$$

$$\therefore \quad m = \left(\frac{nh}{2\pi\, mr}\right)^2 = \frac{Ze^2}{4\pi \in_0 r}$$

$$r = \frac{\in_0 n^2h^2}{\pi\, mZe^2} \quad \text{...(iii)}$$

(ii) $$r = \frac{\in_0 n^2h^2}{\pi\, mZe^2}$$

Here $$m = 200\, m_e$$

$$Z = 3$$

$$r = \frac{\in_0 n^2h^2}{\pi \times 200 m_e \times 3e^2}$$

$$r = \frac{n^2}{600}\left[\frac{\in_0 h^2}{\pi\, m_e\, e^2}\right]$$

For hydrogen atom, for the first Bohr orbit

$$r_H = \left(\frac{\in_0 h^2}{\pi\, m_e\, e^4}\right) = 0.53 \times 10^{-10}\, m$$

$$r = \left(\frac{n^2}{600}\right) r_H$$

But $r = r_H$ (given)

$$\therefore \quad \frac{n^2}{600} = 1,\ n^2 = 600$$

$$n = 24.$$

(iii) $$h\nu = \frac{me^4Z^2}{8\in_0^2 h^2}\left(\frac{1}{n_1^2} - \frac{1}{n_2^2}\right)$$

Here $n_1 = 1, n_2 = 2$

$$U = \frac{200 m_e e^4 (3^2)}{8 \epsilon_0^2 h^2}\left(\frac{1}{1} - \frac{1}{4}\right)$$

$$U = \left(\frac{1800 \times 3}{4}\right)\left(\frac{m_e e^4}{8 \epsilon_0^2 h^2}\right)$$

$$U = \left(\frac{1800 \times 3}{4}\right)\left[\frac{9.1 \times 10^{-31} \times (1.6 \times 10^{-19})^4}{8 \times (8.85 \times 10^{-12})^2 \times (6.624 \times 10^{-34})^2}\right] J$$

$$U = \left(\frac{1800 \times 3}{4}\right)\left[\frac{9.1 \times 10^{-31} \times (1.6 \times 10^{-19})^4}{8 \times (8.85 \times 10^{-12})^2 \times (6.624 \times 10^{-34})^2 \times 1.6 \times 10^{-19}}\right] eV$$

$U = 1.8234 \times 10^4$ eV

Excitation potential

$V = 1.8234 \times 10^4$ Volts.

Example 3:

An X-ray line of wave length 0.53832 Å is found to be emitted from an X-ray tube with Zinc (Z = 30) target in addition to the characteristic K_α line of Zinc of wavelength 1.43603 Å. If the unknown line is due to an impurity in the target, obtain the atomic number of impurity.

Solution:

Here $\lambda_1 = 1.43603$ Å

$Z_1 = 30$

$\lambda_2 = 0.53832$ Å

$Z_2 = ?$

$$\frac{1}{\lambda_1} \propto Z_1^2$$

and $\frac{1}{\lambda_2} \propto Z_2^2$

$$\therefore \quad \left(\frac{Z_2}{Z_1}\right)^2 = \left(\frac{\lambda_1}{\lambda_2}\right)$$

$$Z_2 = Z_1 \left[\frac{\lambda_1}{\lambda_2}\right]^{1/2}$$

$$Z_2 = 30\left[\frac{1.43603}{0.53832}\right]^{1/2}$$

$$Z_2 = 49.$$

Example 4(a):

Calculate the screening fact or for K shell for copper (Z = 29) if the K shell binding energies in hydrogen and copper are respectively 13.5 eV and 8940 eV.

Solution:

For hydrogen

$$U_H = \frac{me^4 Z^2}{8\epsilon_0^2 h^2}$$

$$Z = 1 \text{ and } n = 1$$

$$U_H = \frac{me^4}{8\epsilon_0^2 h^2} = 13.5 \text{ eV}$$

$$= 13.5 \times 1.6 \times 10^{-19} \text{ J} \qquad ...(i)$$

For copper

$$U_{cu} = \frac{me^4 Z_{eff}^2}{8\epsilon_0^2 h^2} = 8940 \text{ eV}$$

$$= (8940 \times 1.6 \times 10^{-19})\text{J} \qquad ...(ii)$$

$$\therefore \quad (13.5 \times 1.6 \times 10^{-19}) = Z_{eff}^2\ (8940 \times 1.6 \times 10^{-19})$$

$$Z_{eff}^2 = 662.2$$

$$Z_{eff} = 25.73 \approx 26$$

$$Z = 29$$

$$\therefore \quad Z_{eff} = (Z - x)$$

$$26 = (29 - x)$$

$$x = 3.$$

Example 4(b):

Calculate the energy and wavelength of the photon emitted from a tungsten target Z = 74 when electron jumps from n = 3 state to n = 1 state due to a vacancy in n = 1 state. Given Energy of electron in n = 1 state of hydrogen atom = –13.6 eV

Solution:

Here $Z = 74$

For $n = 1$, $Z_{Eff} = 74 - 1 = 73$

$$E_1 = \left(\frac{-13.6 Z_{eff}^2}{n^2}\right) eV$$

$$= \left[\frac{-13.6 \times (73)^2}{1^2}\right] eV$$

$$= -7.2474 \times 10^4 \text{ eV}.$$

For n = 3 state, the electron is subject to an effective nuclear charge that depends on the number of electrons in orbits, n = 1 and n = 2. There are eight electrons in n = 2 state and one electron in n = 1 state as there is one vacancy in n = 1 state.

Therefore $Z_{eff} = (Z - 9)$

$$E_3 = \left[\frac{-13.6 \times (Z - 9)^2}{3^2}\right] eV$$

$$E_3 = \left[\frac{(74 - 9)^2 \times 13.6}{9}\right] eV$$

$$= -\left[\frac{(65)^2 \times 13.6}{9}\right] eV$$

$$= -6384 \text{ eV}.$$

(1) Energy of emitted photon,

$$E = E_3 = E_1$$
$$= -6384 - (-72474)$$
$$= 6.609 \times 10^4 \text{ eV} = 6.609 \times 10^4 \times 6 \times 10^{-19} \text{ J}$$
$$= 1.057 \times 10^{-14} \text{ J.}$$

(2) $E = \dfrac{hc}{\lambda}$

$$\lambda = \frac{hc}{E}$$
$$= \frac{6.624 \times 10^{-34} \times 3 \times 10^8}{1.057 \times 10^{-14}} \text{ m}$$
$$= 1.88 \times 10^{-11} \text{ m}$$
$$= 0.188 \text{ Å.}$$

Example 5:

Calculate the wavelength of K_α line in case of copper atom. Take Moseley law into consideration.

Solution:

Take $R = 1.096776 \times 10^7$ per meter.

Here $Z = 29$

Taking Mosely's Law into consideration

$$Z_{eff} = (Z - 1) = 29 - 1 = 28$$
$$n_1 = 1 \text{ and } n_2 = 2$$
$$\frac{1}{\lambda} = R(Z_{eff})^2 \left[\frac{1}{n_1^2} - \frac{1}{n_2^2}\right] = \frac{3R[Z-1]^2}{4}$$
$$\lambda = \frac{4}{3R(Z-1)^2}$$
$$\lambda = \frac{4}{3 \times 1.096776 \times 10^7 \times (28)^2}$$
$$= 1.55 \times 10^{-10} \text{m}$$
$$= 1.55 \text{ Å.}$$

Example 6:

The measured value of the total kinetic energy of the fission fragments from the thermal neutron fission of U^{92}_{235} *is 196 MeV. Assuming that the respective values of Z and A of the fission fragments as (35, 72) and (57, 162). Calcvulate the distance r between the fragments at the instant of separation. Compare this value with the sum of the radio of the two fragments. Nuclear radi* $R = r_0 A^{1/3}$ *where* $r_0 = 1.2 \times 10^{-15}$*m.*

Solution:

Here $Z_1 = 35, Z_2 = 57$

$A_1 = 72, A_2 = 162$

$$U = \frac{(Z_1 e)(Z_2 e)}{4\pi \epsilon_0 r}$$

$$U = 196 \text{ MeV} = 196 \times 10^6 \times 1.6^6 \times 1.6 \times 10^{-19} \text{ J}$$

$$U = 313.6 \times 10^{-13} \text{ J}$$

$$r = \frac{Z_1 Z_2 e^2}{4\pi \epsilon_0 U}$$

$$r = \frac{35 \times 72 \times (1.6 \times 10^{-19})^2 \times 9 \times 10^9}{313.6 \times 10^{-13}}$$

$$r = 1.466 \times 10^{-14}\text{m} \qquad ...(i)$$

(ii) $R_1 = r_0 A_1^{1/3}$

$R_2 = r_0 A_2^{1/3}$

$$R = R_1 + R_2 = r_0 \left[A_2^{1/3} + A_2^{1/3}\right]$$

$$R = 1.2 \times 10^{-15} [(72)^{1/3} + (162)^{1/3})]$$

$$= 1.2 \times 10^{-15} [4.16 + 5.45]$$

$$= 1.27 \times 10^{-14}\text{m}$$

Therefore r is greater than R.

Example 7(a):

Find the energy (in eV) and wavelength of K_α - X-ray of Co-27.

Solution:

$$R = 1.09678 \times 10^7/\text{m}$$

For Co, $Z = 27$

$$Z_{eff} = (Z - 1) = 26$$

$$\bar{v} = (Z-1)^2 R\left[\frac{1}{n_1^2} - \frac{1}{n_2^2}\right]$$

$$\bar{v} = (26)^2 \times 1.09678 \times 10^7 \left[\frac{1}{1} - \frac{1}{4}\right]$$

$$\bar{v} = 5.56 \times 10^9 \text{ per metre}$$

$$\frac{1}{\lambda} = 5.56 \times 10^9 \text{ per metre}$$

(i) $\lambda = 1.799 \times 10^{-10}$ m

$$\lambda = 1.799 \text{ Å}$$

Energy $= \dfrac{hc}{\lambda}$

$$= \frac{6.624 \times 10^{-34} \times 3 \times 10^8}{1.799 \times 10^{-10}}$$

$$= \frac{6.624 \times 10^{-34} \times 3 \times 10^8}{1.799 \times 10^{-10} \times 1.6 \times 10^{-19}} \text{ eV}$$

$$= 6.9 \times 10^3 \text{ eV.}$$

Example 7(b):

Calculate the wavelength of the K_α line from a copper target using Moseley's law. Rydberg constant for hydrogen = 1.09687×10^7 per metre.

Solution:

For Copper $Z = 29$

$$Z_{eff} = (Z - 1) = 28$$

$$\bar{v} = (Z-1)^2 R \left[\frac{1}{n_1^2} - \frac{1}{n_2^2}\right]$$

$$\bar{v} = (28)^2 \times 1.09678 \times 10^7 \left[\frac{1}{1} - \frac{1}{4}\right]$$

$$\bar{v} = 6.449 \times 10^9 \text{ per metre}$$

$$\frac{1}{\lambda} = 6.449 \times 10^9 \text{ per metre}$$

$$\lambda = 1.55 \times 10^{-10}\text{m}$$

$$\lambda = 1.55 \text{ Å}.$$

Example 7(c):

The experimental value of Bohr magneton is 9.21×10^{-24} SI units and Planck's constant $h = 6.6 \times 10^{-34}$ joule-second.

Calcuate the value of e/m of an electron.

Solution:

Here $\beta = 9.21 \times 10^{-24}$ units

$h = 6.6 \times 10^{-34}$ joule-second

$$\beta = \frac{eh}{4\pi m}$$

or
$$\frac{e}{m} = \frac{4\pi\beta}{h} = \frac{4 \times 22 \times 9.21 \times 10^{-24}}{7 \times 6.6 \times 10^{-34}}$$

$$\frac{e}{m} = 1.755 \times 10^{11} \text{ C/kg}.$$

Example 7(d):

What is Bohr Magneton? Calculate its value.

Solution:

Bohr magneton

$$\beta = \frac{eh}{4\pi m}$$

Here $e = 1.6 \times 10^{-19}$ C

$h = 6.624 \times 10^{-34}$ J-s

$m = 9.1 \times 10^{-31}$ kg

$$\beta = \frac{1.6 \times 10^{-19} \times 6.624 \times 10^{-34}}{4 \times 3.142 \times 9.1 \times 10-31}$$

$\beta = 9.27 \times 10\text{–}24$ SI Units.

Example 8:

Calculate the wavelength separation between the two component lines which are observed in the normal Zeeman effect. The magnetic field used is 0.4 weber/m^2, the specific charge = 1.76 × 10^{11} coulombs/kg and λ = 6000 Å.

Solution:

Here $\lambda = 6000$ Å $= 6000 \times 10^{-10}$m

or $\lambda = 6 \times 10^{-7}$m

$$\frac{e}{m} = 1.76 \times 10^{11} \text{ coulombs/kg}$$

$B = 0.4$ weber/m^2

$c = 3 \times 10^8$ m/s

The wavelength separation,

$$\lambda_1 - \lambda_2 = 2\,(d\lambda) = \frac{2Be\lambda^2}{4\pi mc}$$

$$\lambda_1 - \lambda_2 = \frac{Be\lambda^2}{4\pi mc}$$

$$\lambda_1 - \lambda_2 = \frac{0.4 \times 1.76 \times 10^{11} \times (6 \times 10^{-7})^2}{2 \times 3.14 \times 3 \times 10^8}$$

$\lambda_1 - \lambda_2 = 1.335 \times 10^{-11}$m $= 0.1335$ Å.

Example 9:

A 1 MeV proton is sent aganist a gold leaf (Z = 79). Calculate the distance of the closest approach for aq head on collison.

Solution:

$$d = \frac{Ze^2}{4\pi \epsilon_0 \left(\frac{1}{2}\right) m\upsilon^2} = \frac{Ze^2}{(4\pi \epsilon_0)U}$$

$U = 1 \text{ MeV} = 10^6 \times 1.6 \times 10^{-19} \text{ J} = 1.6 \times 10^{-13} \text{ J}$

$e = 1.6 \times 10^{-19}$ C

$Z = 79$

$$d = \frac{79 \times (1.6 \times 10^{-19}) \times 9 \times 10^9}{1.6 \times 10^{-13}}$$

$d = 1.137 \times 10^{-13}$m.

Example 10:

An α particle of energy 2×10^{-13} J is scattered of an aluminium atom through an angle of 90°. Calculate the distance of closest approach. Atomic number of Al is 13, $e = 1.6 \times 10^{-19}$ coulomb and $4\pi\epsilon_0 = 1 \times 10^{-10}$.

Solution:

$$d = \left(\frac{2Ze^2}{4\pi \varepsilon_0 m\upsilon^2}\right)\left(1 + \operatorname{cosec}\frac{\theta}{2}\right)$$

Here $Z = 13$

$e = 1.6 \times 10^{-19}$ C

$\frac{1}{2} m\upsilon^2 = 2 \times 10^{-13}$ J

$m\upsilon^2 = 4 \times 10^{-31}$ J

$\theta = 90°$

$\operatorname{cosec}\frac{90°}{2} = \sqrt{2} = 1.414$

$4\pi \varepsilon_0 = 1 \times 10^{-10}$ (given)

$$d = \left[\frac{2 \times 13 \times (1.6 \times 10^{-19})^2}{10^{-10} \times 4 \times 10^{-13}}\right](1 + 1.414)$$

$$d = 4.169 \times 10^{-14} \text{ m.}$$

Example 11(a):

The Zeeman components of a 5461 Å spectral line are 0.417 Å apart when the magnetic field is 1.5 T. From these data, calculate the value of e/m of an electron.

Solution:

$$\lambda_1 - \lambda_2 = \frac{Be\lambda^2}{2\pi mc}$$

Here $\lambda_1 - \lambda_2 = 0.417 \text{ Å} = 0.417 \times 10^{-10} \text{ m}$

$\lambda = 5461 \text{ Å} = 5461 \times 10^{-10} \text{ m}$

$B = 1.5T$

$C = 3 \times 10^8 \text{ m/s}$

$$\therefore \quad \frac{e}{m} = \frac{(\lambda_1 - \lambda_2)\, 2\pi c}{B\lambda^2}$$

$$\frac{e}{m} = \frac{(0.417 \times 10^{-10}) \times 2 \times 3.14 \times 3 \times 10^8}{1.5 \times (5461 \times 10^{-10})^2}$$

$$\frac{e}{m} = 1.756 \times 10^{11} \text{ C/kg.}$$

Example 11(b):

In observing the Raman spectrum of a sample, using 2537 Å as the exciting line, one gets stoke line at 2683 Å, Deuce the Raman shift in cm^{-1} units. compute the wavelength in Å for corresponding stokes and antistokes lines if the exciting line is 5461 Å.

Solution:

(i) Here $\upsilon_s = \upsilon_0 = \upsilon_m$

$$\upsilon_m = v_0 = v_s$$

$$= \frac{c}{\lambda_0} - \frac{c}{\lambda_s}$$

$$\frac{c}{\lambda_m} = c\left[\frac{1}{\lambda_0} - \frac{1}{\lambda_s}\right]$$

$$\frac{1}{\lambda_m} = \frac{\lambda_s - \lambda_0}{\lambda_0 \lambda_s}$$

Here $\lambda_0 = 2537 \text{ Å} = 2537 \times 10^{-10}$ m

$\lambda_s = 2683 \text{ Å} = 2683 \times 10^{-10}$ m

$$\frac{1}{\lambda_m} = \frac{2683 \times 10^{-10} - 2537 \times 10^{-10}}{2537 \times 10^{-10} \times 2683 \times 10^{-10}}$$

$$\frac{1}{\lambda_m} = 2.145 \times 10^5 \text{ per metre}$$

Raman Shift,

$$\upsilon_m = \frac{1}{\lambda_m} = 2.145 \times 10^5 \text{ per metre}$$

$$= 2.145 \times 10^3 \text{ per cm}$$

(i) Here $\lambda_0 = 5661 \text{ Å} = 5461 \times 10^{-10}$ metre

$$\frac{1}{\lambda_m} = 2.145 \times 10^5 \text{ per metre}$$

$$\lambda_s = ?$$

$$\upsilon_s = \upsilon_0 - \upsilon_m$$

$$\frac{c}{\lambda_s} = \frac{c}{\lambda_0} - \frac{c}{\lambda_m}$$

$$\frac{1}{\lambda_s} = \frac{1}{\lambda_0} - \frac{1}{\lambda_m}$$

$$\frac{1}{\lambda_s} = \left(\frac{1}{5461 \times 10^{-10}}\right) - 2.145 \times 10^5$$

$$\frac{1}{\lambda_s} = 18.31 \times 10^5 - 2.145 \times 10^5$$

$$= 16.165 \times 10^5$$

$$\lambda_s = \frac{1}{16.165 \times 10^5}$$

$$= 6186 \times 10^{-10} \text{m}$$

$$= 6186 \text{ Å}.$$

(ii) Similarly, for antistokes line

$$\upsilon_{as} = \upsilon_0 + \upsilon_m$$

$$\frac{c}{\lambda_{as}} = \frac{c}{\lambda_0} + \frac{c}{\lambda_m}$$

$$\frac{1}{\lambda_{as}} = \frac{1}{\lambda_0} + \frac{1}{\lambda_m}$$

$$\frac{1}{\lambda_{as}} = \left(\frac{1}{5461 \times 10^{-10}}\right) + 2.145 \times 10^5$$

$$\frac{1}{\lambda_{as}} = 18.31 \times 10^5 + 2.145 \times 10^5$$

$$\frac{1}{\lambda_{as}} = 20.455 \times 10^5$$

$$\lambda_{as} = \frac{1}{20.455 \times 10^5}$$

$$= 4889 \times 10^{-10} \text{m}$$

$$= 4889 \text{ Å}.$$

Example 12:

In an experiment with benzene the wavelengths of a pair of Stokes and anti-Stokes lines were measured to be 4554 Å and 4178 Å respectively. Calculate the wavelength of the corresponding infra-red absorption line.

Solution:

Here $\upsilon = \upsilon_0 \pm \upsilon_m$

$\therefore$ $\upsilon_1 = \upsilon_0 + \upsilon_m$

and $\upsilon_2 = \upsilon_0 - \upsilon_m$

$\therefore$ $\upsilon_1 - \upsilon_2 = 2\upsilon_m$

$$\frac{c}{\lambda_1} - \frac{c}{\lambda_2} = \frac{2c}{\lambda_m}$$

$$\lambda_m = \left[\frac{2\lambda_1\lambda_2}{\lambda_2 - \lambda_1}\right]$$

Here $\lambda_2 = 4554$ Å

$\lambda_1 = 4178$ Å

$$\lambda_m = \frac{2 \times 4554 \times 4178}{4554 - 4178}$$

$$\lambda_m = 1.012 \times 10^5 \text{ Å}.$$

Example 13:

Calculate the frequency of Oscillation of a hydrogen molecule if its force constant is 4.8 × 10^2 N/m and mass of hydrogen atom = 1.67 × 10^{-27} kg.

Solution:

$$\upsilon = \frac{1}{2\pi}\sqrt{\frac{K}{\mu}}$$

Here μ is the reduced mass of the system

$$\frac{1}{\mu} = \frac{1}{m} + \frac{1}{m} = \frac{2}{m}$$

$$\mu = \left(\frac{m}{2}\right)$$

Here $K = 4.8 \times 10^2$ N/m

$$\mu = \frac{1.67 \times 10^{-27}}{2} = 0.835 \times 10^{-27} \text{ kg}$$

$$\therefore \quad \upsilon = \frac{1}{2\pi}\sqrt{\frac{4.8 \times 10^2}{0.835 \times 10^{27}}}$$

$$\upsilon = 1.2 \times 10^{-14} \text{ Hz}.$$

Example 14:

Find the radius and speed of the electron in the first bohr orbit of the hydrogen atom. How will the radius and speed of electron change with the increase in atomic number of the atom.

Solution:

(i) The radius of the Bohr orbit is given by

$$r = \frac{\epsilon_0 n^2 h^2}{\pi m Z e^2}$$

For hydrogen,

$$Z = 1, n = 1$$

$$r_H = \frac{\epsilon_0 h^2}{\pi m e^2}$$

Here $\epsilon_0 = 8.85 \times 10^{-12}\ C^2/N - m^2$

$m = 9.1 \times 10^{-31}$ kg

$e = 1.6 \times 10^{-19}$ C

$h = 6.624 \times 10^{-34}$ J-s

$$\therefore \quad r_H = \frac{8.85 \times 10^{-12} \times 6.624 \times 10^{-34})^2}{\pi \times 9.1 \times 10^{-31} \times (1.6 \times 10^{-19^2})^2}$$

$$r_H = 5.30 \times 10^{-11} \text{ m.}$$

Also $r = \frac{r_H}{Z}$

Therefore, radius of the orbit will decrease with increase in atomic number Z.

(ii) velocity of the electron

$$\upsilon = \frac{Ze^2}{2 \epsilon_0 nh}$$

In hydrogen Z = 1, n = 1

$$\upsilon_H = \frac{Ze^2}{2 \epsilon_0 h}$$

$$= \frac{(1.6 \times 10^{-19})^2}{2 \times 8.85 \times 10^{-12} \times 6.624 \times 10^{-34}}$$

$$= 2.2 \times 10^6 \text{ m/s}$$

Also $\upsilon = Z\upsilon_H$

Therefore velocity of the electron in the orbit ill increase ith increase in atomic number Z.

Example 15:

The ionization potential of atomic hydrogen is 13.6 V. Calculate the wavelength of light emitted in a transition starting at the first excited state of hydrogen atom?

Solution:

Ionization potential = 13.6 V

Energy of electron in the first orbit

$$U_1 = -13.6 \text{ eV}$$

Energy of electron in the second orbit

$$U_2 = \frac{-13.6}{n^2} = \frac{-13.6}{4} = -3.4 \text{ eV}$$

$$U_2 - U_1 = -3.4 + 13.6$$

$$h\nu = 10.2 \text{ eV}$$

$$\frac{hc}{\lambda} = 10.2 \times 1.6 \times 10^{-19} \text{ J}$$

$$= 1.632 \times 10^{-18} \text{ J}$$

$$\lambda = \frac{hc}{1.632 \times 10^{-18}}$$

$$= \frac{6.624 \times 10^{-34} \times 3 \times 10^8}{1.632 \times 10^{-18}}$$

$$= 1.217 \times 10^{-7} \text{ m}$$

$$= 1217 \text{ Å}.$$

Example 16:

The wavelength of the first number of Balmer series of hydrogen is

$$6563 \times 10^{-10} \text{ m}$$

Calculate the wavelength of its second number

Solution:

For the first number

$$\bar{v}_1 = R\left[\frac{1}{2^2} - \frac{1}{3^2}\right]$$

$$\bar{v}_1 = \frac{5}{36}R$$

or $$\frac{1}{\lambda_1} = \frac{5}{36}R \qquad ...(i)$$

For the second number,

$$\bar{v}_2 = R\left[\frac{1}{2^2} - \frac{1}{4^2}\right]$$

or $$\frac{1}{\lambda_2} = \frac{3}{16}R \qquad ...(ii)$$

Dividing (i) by (ii)

$$\frac{\lambda_2}{\lambda_1} = \frac{20}{27}$$

$$\lambda_2 = \frac{20}{27}\lambda_1$$

But $$\lambda_1 = 6563 \times 10^{-10} \text{ m}$$

$\therefore$ $$\lambda_2 = \frac{20 \times 6563 \times 10^{-10}}{27}$$

$$= 4861 \times 10^{-10} \text{ m.}$$

Example 17(a):

A beam of electrons is used to bombard gaseous hydrogen. What is the minimum energy in **electron-volts** *the electrons must have if the first number of the Balmer of the Balmer series corresponding to a transition from $n_2 = 3$ state to $n_1 = 2$ state is to be emitted?*

$$h = 6.6 \times 10^{-34} \text{ joule second.}$$

Solution:

$$\text{Energy required} = \frac{me^4}{8\epsilon_0^2 h^2}\left[\frac{1}{n_1^2} - \frac{1}{n_2^2}\right]$$

Here $n_1 = 2, n_2 = 3$

$$\therefore \text{ Energy required} = \left(\frac{me^4}{8 \epsilon_0^2 h^2}\right) \times \left(\frac{5}{36}\right) \text{ joule}$$

But $1 \text{ eV} = 1.6 \times 10^{-19}$ joule

$$\therefore \text{ Energy required} = \frac{5\, me^4}{8 \times 36 \times \epsilon_0^2 h^2 \times (1.6 \times 10^{-19})} \text{ electron-volts}$$

$$= \frac{5 \times 9 \times 10^{-31} \times (1.6 \times 10^{-19})^4}{8 \times 36\, (8.85 \times 10^{-12}) \times (6.6 \times 10^{-34})^2 \times (1.6 \times 10^{-19})}$$

$= 1.88$ eV.

Example 17(b):

Calculate the ionization potential in electron volts for hydrogen atom given that

$e = 1.6 \times 10^{-19}$ coulomb

$m = 9 \times 10^{-31}$ kg

$h = 6.6 \times 10^{-34}$ joule-second

$\epsilon_0 = 8.85 \times 10^{-19}$ coulomb²/newton – m².

Solution:

Work function,

$$\phi = \frac{me^4}{8 \epsilon_0^2 h^2} \text{ joules}$$

$$\phi = \frac{9 \times 10^{-31} \times (1.6 \times 10^{-19})^4}{8 \times (8.85 \times 10^{-12})^2 \times (6.6 \times 10^{-34})^2} \text{ joules}$$

But $1 \text{ eV} = 1.6 \times 10^{-10}$ joule

$$\therefore \quad \phi = \frac{9 \times 10^{-31} \times (1.6 \times 10^{-19})^4}{8(8.85 \times 10^{-19})^2 \times (6.6 \times 10^{-34}) \times 1.6 \times 10^{-19}}$$

or $\phi = 13.51$ eV

Example 18:

The wavelength of the second line of the Balmer series in the hydrogen spectrum is 4861 Å. Calculate the wavelength of the first line.

Solution:

For the second line,

$$\bar{v}_2 = R\left[\frac{1}{2^2} - \frac{1}{4^2}\right] \bar{v}_2 = \frac{3}{16}R$$

$$\frac{1}{\lambda_2} = \frac{3}{16}R \qquad \text{...(i)}$$

For the first line,

$$\bar{v}_1 = R\left[\frac{1}{2^2} - \frac{1}{3^2}\right]$$

$$\bar{v}_1 = \frac{5}{36}R$$

or $$\frac{1}{\lambda_1} = \frac{5}{36}R \qquad \text{...(ii)}$$

Dividing (i) by (ii)

$$\frac{\lambda_1}{\lambda_2} = \frac{27}{20}$$

$\therefore$ $$\lambda_1 = \frac{27\,\lambda_2}{20}$$

But $$\lambda_2 = 4861 \text{Å}$$

$$\lambda_1 = \frac{27 \times 4861}{10}$$

or $$\lambda_1 = 6563 \text{ Å}.$$

Example 19(a):

The wavelength of sodium D_1 line is 590 nm. Calculate the difference in energy levels involved in the emission or absorption of this line.

Solution:

Here $E_{(n_2-n_1)} = h\nu = \frac{hc}{\lambda}$

Here $h = 6.624 \times 10^{-34}$ J-s

$c = 3 \times 10^{8}$ m/s

$h = 590 \text{ nm} = 590 \times 10^{-9}$ m

$$\therefore \quad E_{(n_2-n_1)} = \frac{(6.624 \times 10^{-34}) \times 3 \times 10^{8}}{590 \times 10^{-9}}$$

$$= 3.37 \times 10^{-19} \text{ J.}$$

Example 19(b):

Wavelength of Balmer H_α line is 6563 Å. Calculate the wavelength of H_β line.

Solution:

For H_α line of Balmer series *i.e.,* the second number,

$$\bar{\nu}_1 = R\left[\frac{1}{2^2} - \frac{1}{3^2}\right]$$

$$\frac{1}{\lambda_1} = \frac{5}{36}R \qquad \text{...(i)}$$

For H_β line of Balmer Series *i.e.,* the second number,

$$\bar{\nu}_2 = R\left[\frac{1}{2^2} - \frac{1}{4^2}\right]$$

$$\frac{1}{\lambda_1} = \frac{3}{16}R \qquad \text{...(ii)}$$

Dividing (i) by (ii)

$$\frac{\lambda_2}{\lambda_1} = \frac{20}{27}$$

$$\lambda_2 = \left(\frac{20}{27}\right)\lambda_1$$

But $\lambda_1 = 6563$ Å

$$\lambda_2 = \left(\frac{20}{27}\right) 6563$$

$$\lambda_2 = 4861 \text{ Å}.$$

Example 19(c):

Calculate the energy required to excite the hydrogen atom from the ground state (n = 1) to the first excited state (n = 2).

Solution:

Energy required $U = \frac{me^4}{8 \epsilon_0^2 h^2}\left[\frac{1}{n_1^2} - \frac{1}{n_2^2}\right]$

$$U = \frac{me^4}{8 \epsilon_0^2 h^2}\left[\frac{1}{1^2} - \frac{1}{2^2}\right]$$

$$U = \frac{3me^4}{32 \epsilon_0^2 h^2}$$

Here $m = 9.1 \times 10^{-31}$ Kg, $e = 1.6 \times 10^{-19}$ C

$\epsilon_0 = 8.85 \times 10^{-12}$ C²/N – m²

$h = 6.624 \times 10^{-34}$ J-s

$$U = \frac{3 \times 9.1 \times 10^{-31} \times (1.6 \times 10^{-19})^4}{32 \times (8.85 \times 10^{-12})^2 \times (6.624 \times 10^{-34})^2} \text{ joules}$$

$$U = \frac{3 \times 9.1 \times 10^{-31} \times (1.6 \times 10^{-19})^4}{32 \times (8.85 \times 10^{-12})^2 \times (6.624 \times 10^{-34})^2 \times 1.6 \times 10^{-19}} \text{ eV}$$

$$U = 10.13 \text{ eV}.$$

Example 20:

Calculate the radius of the first Bohr orbit for (i) H and (ii) He atoms and also the velocity of the electron in these orbits as compared to the velocity of light.

Solution:

For hydrogen atom, the radius of the orbit is

$$r_H = \frac{\epsilon_0 n^2h^2}{\pi mZe^2}$$

Here $n = 1, Z = 1$

$\epsilon_0 = 8.85 \times 10^{-12}\ C^2/N - m^2$

$m = 9.1 \times 10^{-31}$ kg

$e = 1.6 \times 10^{-19}$C

$h = 6.624 \times 10^{-34}$ J-s

$$\therefore \quad r_H = \frac{8.85 \times 10^{-12} \times 1 \times (6.624 \times 10^{-34})^2}{3.142 \times 9.1 \times 10^{-31} \times (1.6 \times 10^{-19})^2}$$

$r_H = 5.30 \times 10^{-11}$ m.

For helium, atom, $n = 1, Z = 2$

$$r_H = \frac{r_H}{Z} = \frac{5.30 \times 10^{-11}}{2}$$

$= 2.65 \times 10^{-11}$m.

(ii) velocity of the electron,

$$\upsilon = \frac{Ze^2}{2\epsilon_0 nh}$$

In hydrogen $Z = 1$, and $n = 1$

$$\upsilon_H = \frac{e^2}{2\epsilon_0 h}$$

$$= \frac{(1.6 \times 10^{-19})^2}{2 \times 8.85 \times 10^{-12} \times 6.624 \times 10^{-34}}$$

$= 2.2 \times 10^6$ m/s

Also, $$\frac{\upsilon_H}{c} = \frac{2.2 \times 10^6}{3 \times 10^8} = 7.33 \times 10^{-3}$$

For helium $$\upsilon_{He} = \frac{Ze^2}{2\epsilon_0 nh}$$

Here $n = 1, Z = 2$

$$\upsilon_{He} = \frac{2e^2}{2 \in_0 h} = 2\upsilon_H$$

$$\upsilon_{He} = 2 \times 2.2 \times 10^6$$

$$= 4.4 \times 10^6 \text{ m/s}$$

$$\frac{\upsilon_{He}}{c} = \frac{4.4 \times 10^6}{3 \times 10^8}$$

$$= 1.466 \times 10^{-2}.$$

Example 21(a):

In hydrogen atom the electron is replaced by a muon whose mass is 200 times that of an electron and charge is same as that of electron, calculate the ionization potential on the basis of Bohr's theory.

Solution:

In the case of hydrogen atom, having an electron, the ionization potential

$$\phi = \frac{me^4}{8 \in_0^2 h^2} \qquad ...(i)$$

when electron is replaced by muon.

$$m_1 = 200 \text{ m}$$

$$\phi_1 = \frac{m_1 e^4}{8 \in_0^2 h^2} = \frac{(200 m) e^4}{8 \in_0^2 h^2} \qquad ...(ii)$$

Dividing (ii) by (i)

$$\frac{\phi_1}{\phi} = 200$$

But $\phi = 13.6$ eV

$$\therefore \qquad \phi_1 = 200 \times 13.6$$

$$= 2.72 \times 10^3 \text{ eV}.$$

Example 21(b):

Calculate the difference in wavelength in the spectra of hydrogen and heavy hydrogen corresponding to the first line on the long wave side of Balmer series.

$$\begin{bmatrix} R_H = 1.097 \times 10^7/\text{m} \\ \text{m} = 0.000549\ M_H \end{bmatrix}$$

Solution:

$$R_H = \frac{R}{1 + \dfrac{m}{M_H}} \qquad \text{...(i)}$$

$$R_D = \frac{R}{1 + \dfrac{m}{M_D}}$$

$$\frac{m}{M_H} = 0.000549$$

$$\frac{m}{M_D} = \frac{m}{2M_H} = 0.000274$$

$$\frac{R_D}{R_H} = \frac{1 + \dfrac{M}{M_H}}{1 + \dfrac{m}{M_D}} = \frac{1 + 0.000549}{1 + 0.000274}$$

$$\frac{R_D}{R_H} = 1.00027$$

$$R_D = (R_H)\ 1.00027$$

$$R_D = 1.097 \times 10^7/\text{m}$$

In the case of Balmer series, for the first member

$$n_1 = 2,\ n_2 = 3$$

$$\frac{1}{\lambda_H} = R_H \left[\frac{1}{n_1^2} - \frac{1}{n_2^2}\right] = \left(\frac{5}{36}\right) R_H$$

$$\lambda_H = \frac{36}{5R_H} = \frac{36}{5 \times 1.097 \times 10^7}$$

$$\lambda_H = 6.5633 \times 10^{-7}\ \text{m}$$

Similarly,

$$\frac{1}{\lambda_D} R_D \left[\frac{1}{n_1^2} - \frac{1}{n_2^2}\right] = \left(\frac{5}{36}\right) R_D$$

$$\lambda_D = \frac{36}{5R_D}$$

$$\lambda_D \frac{36}{5 \times 1.0973 \times 10^7}$$

$$\lambda_D = 6.5615 \times 10^{-7} \text{m}$$

$$\therefore \quad \Delta\lambda = \lambda_H - \lambda_D$$

$$= 6.5633 \times 10^{-7} - 6.5615 \times 10^{-7}$$

$$= 0.0018 \times 10^{-7} \text{m}$$

$$= 1.8 \text{ Å}.$$

Example 22:

What is the energy, momentum and wavelength of the photon emitted by a hydrogen atom when an electron makes a transition from n = 2 to n = 1 state. Given ionization potential = 13.6 eV.

Solution:

Energy of electron in the first orbit of hydrogen atom,

$$E_1 = 13.6 \text{ eV}$$

Energy of electron in second orbit

$$E_2 = \frac{E_1}{n^2} = \frac{E_1}{4} = \frac{-13.6}{4}$$

$$E_2 = -3.4 \text{ eV}.$$

(1) Energy of photon emitted = $E_2 - E_1$

$$E = -3.4 + (13.6)$$

$$= 10.2 \text{ eV}$$

$$= 10.2 \times 1.6 \times 10^{-19} \text{ J}$$

$$= 16.32 \times 10^{-19} \text{ J}$$

(2) Momentum, $p = \frac{E}{c} = \frac{16.32 \times 10^{-19}}{3 \times 10^8}$

$$p = 5.44 \times 10^{-27} \text{ kg} - \text{m/s}.$$

(3) Wavelength

$$\lambda = \frac{h}{p}$$

$$= \frac{6.624 \times 10^{-34}}{5.44 \times 10^{-27}}$$

$$= 1.218 \times 10^{-7}\text{m}$$

$$= 1218 \text{ Å}.$$

EXERCISES

1. What do you understand by energy level diagram. Compare the energy levels of hydrogen with those of singly ionized Helium.
2. Calculate the velocity of the electron in the first orbit of the hydrogen atom.
3. State and explain Bohr's postulates regarding the structure of the hydrogen atom. Derive an expression for the energy of the electron in the first orbit is (1/137)c, where c is the velocity of light.
4. Assuming p = b/2 cot (φ/2) obtain (for Rutherford theory of a α-particle scattering) an expression for the number of particles scattered on the a unit area of a screen at a given distance from the foil at a given angle f from the original direction.
5. What are the main assumptions of Rutherfords's α-particle scattering formula? Assuming p = b/2 cot (φ/2) obtain an expression for the number of particles scattering on to a unit area of a screen at a given angle from the foil at a given angle from the original direction.
6. Give an account of Bohr-Sommerfeld model of elliptical electron orbits of an atom and show that the total energy of an electron moving in Sommerfeld orbit of the same quantum number is the same and identical with that of the corresponding Bohr orbit.
7. Describe and explain Stern-Gerlach experiment and indicate its importance in atomic physics.
8. Give an account of various quantum numbers used to specify completely the state of an electron in an atom.

9. State and explain Pauli's Exclusion Principle. Apply it to account for the periodic classification of elements.
10. Show that the quantum and classical concepts of the hydrogen atom make identical predictions in the limit of large quantum numbers.
11. Give an account of normal Zeeman Effect.
12. The wavelength of the second member of Balmer series of the hydrogen is 4861 Å. Calculate the wavelength of its first number.
13. A beam of electrons is used to bombard gaseous hydrogen. What is the minimum energy (in electron volts) the electrons must have if the first member of the Balmer Series corresponding to a transition from n = 2 state n = 1 state.
14. The experimental value of Bohr magneton is 9.21×10^{-24} SI units. Calculate the specific charge of the electron.

2

Nuclear Emulsion

INTRODUCTION

As charged particles pass through a photographic emulsion of a photographic plate they interact with the tiny silver halide crystals producing a line of latent images along their path. In 1911, M Reinganum in Germany recognized that individual α-particles traversing a photographic emulsion produced separate tracks in a photographic emulsion as a series or closely spaced black specks of silver. Because of the relatively high density and stopping power of the photographic emulsion, the tracks produced by charged particles were very short compared with those obtained in a cloud chamber but do not differ greatly from bubble tracks. Miss Blau and Wambacher, in 1937, discovered 'Star' due to the interaction of high energy cosmic rays with the atomic nucleus in the emulsion. The discovery of radioactivity by Becquerel (1896) can be explained as: The optical photographic emulsions is not suitable for quantitative work with nuclear radiations. The sensitivity is low and the tracks due to charged particles have non-clear range because the developed crystal grains are large and widely spaced.

The composition of the emulsion was changed so as to make it more suitable for study of various ionizing particles, such as alpha particles, protons, mesons and even electrons The nuclear emulsions differ from the optical emulsions in that they have considerably higher silver halide content and smaller grain size In nuclear emulsion the thickness is greater than that of optical emulsion. The smaller the grain size, the more sensitive the emulsion to ionizing particles. Thus different commercially available emulsions, differing chiefly in grain size, can be used to discriminate between different particles.

PULSE AMPLIFIER

In nuclear radiation detectors, *e.g.*, in an ionisation chamber, the pulse height is very small and is of the order of micro volts. Since output voltages of 10 to 100 volts are often required for electronic pulse height discrimination, the gain requirement might be 10^5 to 10^6 for the amplifier. If we arc only interested in counting the ionization pulses then it is not necessary for the height of the output of the amplifier to be accurately proportional to the height of the counter pulse. In order to determine the energy of the ionizing particle, we require linear amplifier. To maintain proportionality between input pulse height and output, the stability of the amplifier is important. Linearity is obtained by proper choice of operating points for the amplification elements and by the use of negative feedback.

The pulse amplifier usually consists of three units in cascade. The first unit is the pre-amplifier which is attached to the nuclear detector by a very short cable This reduces distributed capacitance, which decreases rise time. Its gain is usually about 1 to 100. The output stage of the pre-amplifier is a cathode follower with its characteristic low output impedance.

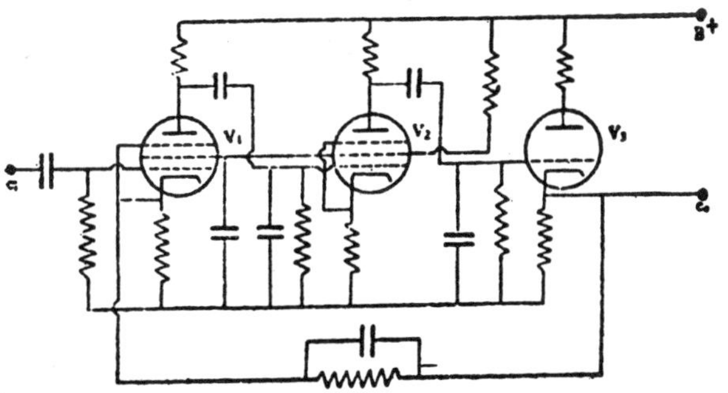

Fig. 2.1 : Linear Pulse Amplifier.

It is, therefore, practical to connect the output terminals of the pre-amplifier by a coaxial screened cable several feet long to the input terminals of the main amplifier, comprising cathode follower and negative feedback. The coupling network CR between pentode valves may be used as the short time constant differentiating network.

The third stage of the unit is a triode connected pentode or a triode with a cathode load resistance shunted by a small trimming capacitor. The output signal is feedback into the input stage through the direct connection between cathode resistances of pentode V_1 and triode V_3. The trimmer capacitor is adjusted so that no transient overshoot occurs with a fast rising input pulse to the amplifier.

DETECTORS OF SOLID-STATE

The first junction type detector was demonstrated by K.G. Mckay in the United States in 1949. A collimated beam of α-particles was allowed to fall on a n-p-germanium junction. He showed that nearly whole of energy of *α-particles absorbed in it was utilized in producing electron hole pairs and the average energy required to produce one electron hole pair was about 3eV which was very small in comparison to 34eV, the energy required to form an ion pair in a gas.* J.D. Mayer and B. Gossick showed that *the size of the output pulse was proportional to the alpha-particle energy. In this respect, a solid state detector behaves like an ideal pulse type ionization chamber.* Most semi-conductor junction type detectors fall into three categories, according to the formation of junction. Crystal counters are essentially the ionization chambers using solid dielectrics between parallel plate electrodes.

The first solid state nuclear detector was given by P.J. Van Heerden in 1945. A crystal of silver chloride (*intrinsic semiconductor*) was placed between two electrodes to which a voltage was applied. When the crystal was exposed to the radiations, the some of the valence band electrons got energy and entered the conduction band, leaving the (*positive*) holes. Under an electric field the electrons moved toward the negative electrode and holes moved toward the positive electrode and a pulse of current equivalent to the number of electrons transferred to the conduction band was produced.

Diffused Junction Detector

In this detector, a thin layer of the n-type is formed on the surface of a p-type silicon (or *germanium*) crystal, by allowing a donor impurity (*phosphorous*) to diffuse into one face of the crystal at a fairly high temperature. There is a tendency for electrons to diffuse from the n-region (*where they are in excess*) into p-region (where the electrons are in deficit).

An equilibrium is reached when the fermi-level is constant throughout the material. Due to this an intermediate volume around the interface will be cleared from carriers of both signs and this region is called the *depletion layer*. This layer is of low conductivity and high field, hence offers ideal conditions for the detection of ionizing radiation.

When the negatively charged electrons diffuse away from the n-type region, they leave the material positively charged. Similarly, the filling of the hole in the p-region generates a corresponding negative charge. The small potential difference of about a volt, established in this manner, is enhanced by an additional potential of several hundred volts, referred to as a reverse-bias voltage.

In the presence of this electric field the positive holes in the p-type side are pulled towards the –ve electrode, the electron towards the p-n junction and thus, a depletion layer is created in the p-type material. The result is an extremely low leakage current. The thickness of the depletion layer increases with the reverse-bias voltage and the limit is set by electrical break-down in the semi-conductor.

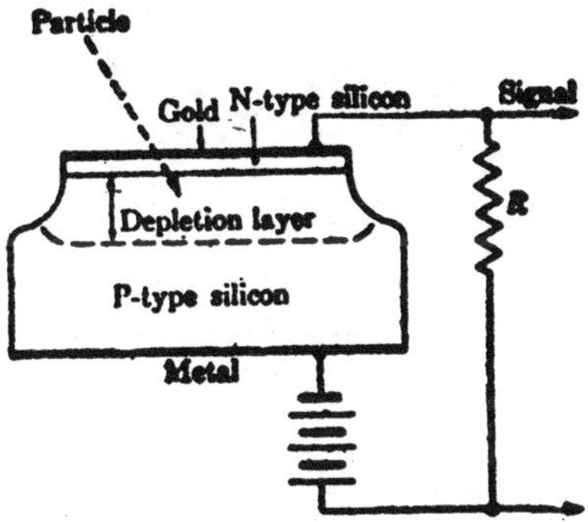

Fig. 2.2 : A p-n junction detector.

Thickness of Depleted Region

The thickness of the depleted region and the capacitance can be calculated by Poisson's equation. Assuming the variation of V in the direction ⊥ to the surface separating the n- and p-regions, we have

$$\frac{d^2V}{dx^2} + \frac{\rho}{\epsilon} = 0 \qquad ...(1)$$

ρ the volume charge density given by

$$\rho = q(p + N_{\&} - n - N_a) \qquad ...(2)$$

where p, n, $N_{\&}$ and N_a denote, respectively, the concentrations of holes, electrons, donors and acceptors. In the completely depleted region n and p are negligible, hence we have

$$\rho = q\,N_{\&} \quad 0 < x < x_1$$

$$\rho = -q\,N_a \quad -x_2 < x < 0$$

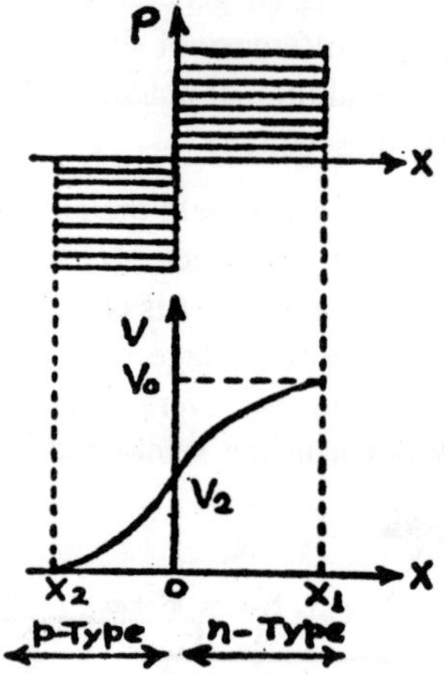

Fig. 2.3 : Charge and potential disturbance.

After successive integrations of eqn. (1), we get

$$V = -\frac{qN_d}{\in}\frac{x^2}{2} + \frac{qN_d}{\in}xx_1 + V_2$$

$$\therefore \qquad V(x_1) = V_0 = qN_d x_1^2/2\in + V_2$$

$$\text{or} \qquad V_0 - V_2 = qN_d x_1^2/2\in \qquad ...(3)$$

Similarly $V(-x_2) = 0$ and $V_2 = qN_a\dfrac{x_2^2}{2\in}$ ...(4)

Hence the potential barrier height becomes

$$V_0 = q(N_d x_1^2 + N_a x_2^2)/2\in \qquad ...(5)$$

As charge is conserved, hence $N_d x_1 = N_a x_2$ and

$$V_e = \frac{q}{2\in}\left(1 + \frac{N_a}{N_d}\right)N_a x_2^2 = \frac{q}{2\in}\left(1 + \frac{N_d}{N_a}\right)N_d x_1^2$$

The thickness of the depletion layer

$$X_0 = x_1 + x_2 = \left[\frac{2 \in V_0}{q}\frac{N_a + N_d}{N_a N_d}\right]^{1/2} \quad ...(6)$$

If p-type region is more heavily doped, than the other region, we have $N_a > N_d$ and $x_2 < x_1$. Hence, we have

$$X_0 \simeq x_1 = \left[\frac{2 \in V_0}{q N_d}\right]^{1/2} = [2 \in \rho_n \mu_n V_0]^{1/2} \quad ...(7)$$

where ρ_n is the resistivity of n-type region and μ_n the electron mobility.

The same treatment will give the value of X_0 in the presence of reverse bias, as

$$X = \left[\frac{2 \in (V_0 + V)}{q N_d}\right]^{1/2} = [2 \in \rho_n \mu_n (V_0 + V)]^{1/2}$$

If the applied voltage $V > V_0$, we get

$$X \simeq [2 \in \rho_n \mu_n V]^{1/2}, \quad ...(8)$$

where a is a constant of a medium.

The depletion layer having area A has a capacitance C given as

$$C = \frac{\in A}{X} = A\left[\frac{q \in N_a N_d}{2(N_a + N_d)(V_a + V)}\right]^{1/2}$$

For $N_a > N_d$ and $V > V_0$, we have

$$C \simeq A (\in / 2 \rho_n \mu_n V)^{1/2}. \quad ...(9)$$

If ρ is in ohm cm, V in volts, X in microns (μ), and C in pf/cm^2 then we have following values:

G_e	n-type	$X = (\rho V)^{1/2}$	$C = 1.37 \times 10^4 (\rho V)^{-1/2}$
	p-type	$= 0.65 (\rho V)^{1/2}$	$= 2.12 \times 10^4 (\rho V)^{-1/2}$
Si	n-type	$X = 0.5 (\rho V)^{1/2}$	$C = 2.2 \times 10^4 (\rho V)^{-1/2}$
	p-type	$= 0.3 (\rho V)^{1/2}$	$= 3.7 \times 10^4 (\rho V)^{-1/2}$

These detectors should always be kept in complete darkness, otherwise the photosensitivity will bring out a *photoelectric current.*

Surface Barrier Detector

It has been very useful for charged particles detection. It is made by exposing a chemically etched surface of an n-type silicon crystal to air. The etched surface is thus oxidised and this oxidation laye-acts like a very thin p-type layer.

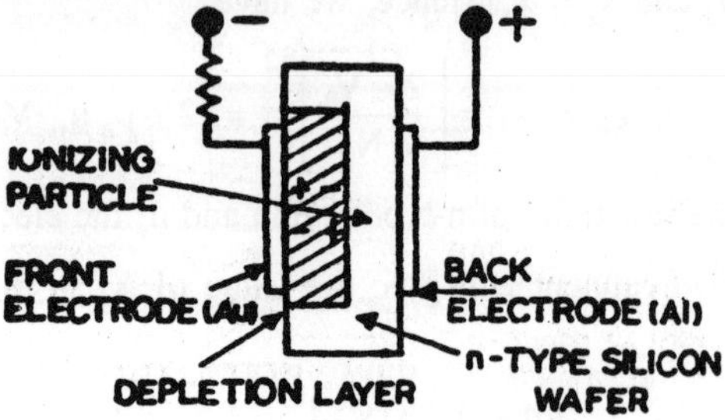

Fig. 2.4 : Surface barrier detector.

In analogy to an N-P junction, a depletion layer is formed, but it is clear that this layer only exists with in the semiconductor. The thickness of the depletion layer X_0 (or X with the reverse bias V) and the capacitance can be written as given in the case of N-P junction. Using 4000Ω silicon with the reverse bias of 1000 volts, this thickness reaches to 3 mm. In contrast to N-P junction, the temperature attained during the various stages of the construction of the surface barrier detectors does not exceed the boiling point of water, and the life time remains as that of the original material.

Detection Characteristics of Surface Barrier Detectors

(a) *Resolution:* The high energy resolution means a narrow distribution of pulse size around the mean, *i.e.,* FWHM (full width at half maxima). With monoergic radiations,, Siffert, *et al.*, found that the value of FWHM was smaller for protons and deuterons than for α-particles. The broadening of peak for α-particles was due to nuclear collisions. The FWHM is found to be ~10 keV for α-particles of several MeV. Mayer and others suggested FWHM of 2.1keV at 135°K. The cooling increases the resolution. It is thus usually necessary to cool the detector to –30°C.

(b) *Linearity:* Many experiments verify the linearity of the response to various particles with different energies. In addition to the defect related to a thin space charge region, other deviations arise for heavy particles and fission fragments.

(c) *High Conversion Efficiency:* The energy required to produce ion pair in silicon is 3.5eV. Thus, the number of pairs for a particular incident energy is about 10 times the number of pairs in air.

(d) *Differential Sensitivity:* The surface barrier detectors are relatively insensitive to neutrons or photons. They can detect charge particles in the presence of other radiations.

(e) *Small Size:* These detectors are of very small size and are important. In some cases the small sizes are a drawback.

(f) *High Speed of Response:* The carrier mobilities are high (~1500cm^{2}volt/sec) for electrons and 500 for holes. The ion pairs move only a short distance before the collection and the pulse widths are in the order of nanoseconds.

Lithium Ion Drifted Junction Detector

The above limitation was reduced by E.M. Pell in the United States in 1960 after introducing the lithium-ion drifted junction detector. The technique is to allow the lithium, which acts as an electron donor, to diffuse into the p-type silicon or germanium at an elevated temperature (120 to 150°C) and under reverse bias.

Under these conditions the ions drift for a considerable distance into the crystal where the lithium donor exactly compensates the existing acceptor impurities in the p-region and an effective intrinsic semi-conductor layer is produced between the n-and p-regions.

Advantage

1. The elimination of windows and the requirement of low applied voltage to operate the counter are advantages over the gas counters.
2. These are more suitable than the grid ionization chambers coincidence experiments.
3. Very good energy resolution is possible.
4. Counting rates high as 5×10^4 counts/sec. are possible.
5. If the particle range is less than the junction thickness, the pulse height is proportional to the K.E. of the incident protons,

deuterons, α-particles or any other ionizing particles.

6. These detectors have low sensitivity to gamma background radiation.

The least sensitive emulsions will show only fission fragment tracks, whereas the most sensitive ones reveal the tracks of singly charged minimum-ionizing particles.

Table 1 : Composition in kg/m³ of Dry Nuclear Emulsions

↓ Element→ Type	H	C	N	O	S	Br	Ag	I
Ilford	49	300	73	200	11	1465	2025	57
Kodak Ltd.,	38	270	80	160	–	1440	1970	36
Eastman Kodak	43	340	110	170	–	1220	1700	54

A series of nuclear emulsions has been developed for detecting particles of different specific ionization. D-l emulsion is the suitable type for detecting fission fragments and alpha particles. The C-2 Ilford emulsion is a useful general purpose emulsion, suitable for recoiling mesons, protons and heavy nuclei.

Table 2 : Nuclear and Optical Emulsions

Property	Optical Emulsion	Nuclear Emulsion
AgBr : Gelatin (mass)	4 : 53	80 : 20
AgBr : Gelatin (volume)	15 : 85	45 : 55
Grain size (micron)	1—3.5	0.1—0.6
Thickness (")	2—3	25—2000
Sensitivity to light	Very high	Poor
Response to α-particles	Dense blackening	Individual tracks
” ” β ”	Moderate ”	Faint fog
” ” γ ”	Faint ”	Almost none

The Ilford G-5 is used to record high energy electrons. The tracks produced in an emulsion are very short. They can be magnified and photographed. High quality microscopes are required for sinning the developed plates. The nuclear emulsion with their high silver halide concentration shrink during the fixation process unfortunately. A knowledge of this shrinkage factor is necessary when we want to determine the angle of dip of a track in the emulsion.

Table 3 : Characteristics of Nuclear Emulsions

Ilford Emulsion type	D1	E-1(K1)	C_2(K2)	B-2	G-5
Mean grain diameter (mm)	0.12	0.14	0.10	0.21	0.18
Highest detectable velocity β	0.2	0.31	0.46	all	
Highest detectable energy of protons in MeV		20	50	120	all

Fast neutrons can be detected by the knock on protons in unloaded emulsions. Slow neutrons can be detected by using emulsions impregnated with lithium, boron or uranium. Gamma rays of energy above about 3 MeV can be detected through the track of the protons, released in photo-disintegration reaction, in emulsions loaded with deuterium compounds.

Under favourable conditions it is possible to determine the charge, mass and incident K.E. of a charged particle brought to rest in a nuclear emulsion. The range of the particle, the grain density and its variation along the track and the small scattering angle due to collisions of the particle with nuclei in the emulsion are all relevant to these determinations. We shall confine ourselves here to discussion of the main relationships which have been established by experiments. The number of grains deposited per micron dn/dx of track is closely related to the energy loss per unit of track dT/dx keV/micron. These two quantities are related by the empirical relation

$$\frac{dn}{dx} = a\left\{1 - e^{bz(dT/dx^{1/2} - c^{1/2})}\right\}, \qquad ...(10)$$

where a, b and c are empirical constants and z is the integer characterizing the charge of the ionizing particle.

If m (the mass in units of proton mass) and z of a particle are known, its kinetic energy T (MeV) may be obtained from its total range R (microns) by an equation

$$T = Kz^{2n}\, m^{1-n}\, R^n, \qquad ...(11)$$

where K and n are empirical constants.

The total number of silver grains deposited and the total particle range are related by an equation

$$N = K' z^{2n'} m^{1-n'} R^{n'}, \quad ...(12)$$

where K' and n' are another constants.

Fig. 2.5 gives the range of protons, deuterons, α-particles, pions and muons in Eastman Kodak NTA emulsions as a function of particle energy.

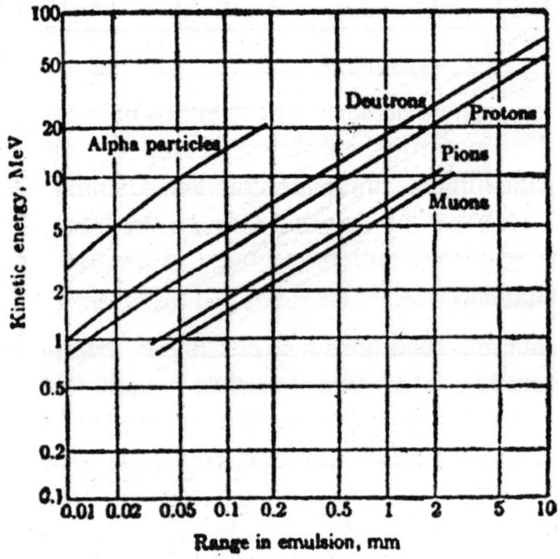

Fig. 2.5 : Range V/s energy for various particles in nuclear emulsions.

As compared with track chambers, like bubble and cloud chambers, photographic emulsions have following *drawback.* It is very difficult to determine the sign of the charge and the momentum of a particle from observations of the curvature of the path in a strong magnetic field.

The actual path length in the emulsion is small compared to that in a cloud chamber and the multiple Coulomb scattering in the emulsion stimulates the curvature. By the use of very strong magnetic fields, the deflection of long (but not short) tracks in emulsions has been observed. This gives the sign of the charge, but the large amount of scattering makes the deflected tracks useless for the momentum calculations.

Some of the *advantages* of the nuclear emulsion detector are:

1. It produces a visible track following the complete trajectory of a charged particle. Particles of different ionizing powers produce tracks which appear quite different.
2. The emulsion is continuously sensitive and is consequently always available to record an event.
3. The energy of product particles can be determined from their range in the emulsion and the differential scattering cross section from counts of the number of tracks formed in each emulsion.
4. The emulsion is relatively light and cheap. This makes it better for high altitude cosmic ray experiments.
5. Due to high stopping power of emulsions, many short lived particles can be brought to rest in the emulsions, before they decay.
6. The measurement of the ranges of muons emitted from stopped pions, before they decay into electrons, enables the mean lifetime of the muon to be found.
7. They were widely employed in cosmic ray studies and led to the discovery of the π- and K-mesons. They are also employed in observations on hyperfragments.

The properties and applications of various nuclear detectors are listed in Table 4.

Table 4 : Common Nuclear Detectors

Detector	Detecting	Particles System	Application Detected
Ionization chamber	Air	α	Measurement of activity
Ionization chamber (with grid)	Argon + CO_2	α, fission fragments	Energy measurement, if particle stops in chamber.
Proportional counters	Argon + CH_4	α, β, electrons	Measurement of energy.
Geiger-Müller counters	Argon + $C/_2$	α, β, γ, cosmic-rays	Good for intensity determination (counts pulses).
Solid-state dete- parti-	Silicon doped	α, β, p, d, t,	Detection of charged
ctors (few mic- rons to a few	with Li, P or Boron	fission products	cles. Measurement of energy.

mm. in thickness)			
Solid-slate detector (few mm. to 10 cm in thickness)	Germanium doped with Li	γ	Energy discrimination, but low efficiency.
Scintillation detector (NaI)	NaI (Thallium activated)	γ, high energy particles	Energy measurement.
Solid-organic; Scintillators	Anthracene stilbene	β^-, p	Energy selection for electrons, Detection of recoil protons.
Liquid-organic from Scintillators	Toluene	β^-, soft X-rays	Detection of soft-rays element dissolved.
Cerenkov counters	Glass, mica, cellophane	Electrons, X-rays, charged particles	Detection of high energy X-rays, To distinguish charged particles.
Cloud chamber	Alcohol +water	Charged particles	Determination of energy, Particle detection.
Bubble chamber energetic	Propane, Xenon, liquid H_2	Charged particles	Detection of very particles. To study particle interaction.
Spark chamber trajectories.	Metallic plates	High energy charged particles	Localization of
Photographic emulsions parti-cle interaction.	Emulsions	Charged particles	Determination of energy and charge. To study

REGIONS OF MULTIPLICATIVE OPERATION

Ion current or pulse height in an ionization chamber increases above the saturation value at sufficiently high applied voltages because electrons moving in these high fields acquire enough energy to cause secondary ionization. The size of the current pulse depends on the number of initial ion pairs produced by the ionizing particles and on the applied voltage. For a given ion pairs, the pulse size is shown in Fig. 2.6 on a logarithmic scale as a function of the applied voltage.

By these results shows that the curve can be divided into six more distinct regions.

In region I, the voltage applied is so low that recombination of ion pairs takes place. As the voltage is increased there is a competition between the loss of ion pairs by recombination and the removal of ions by collection on the electrodes. The region is known as ion recombination region.

In region II, the voltage is sufficiently high so that only a negligible amount of recombination takes place and the ion pairs move to the electrodes so rapidly that virtually every ion pair reaches the electrodes.

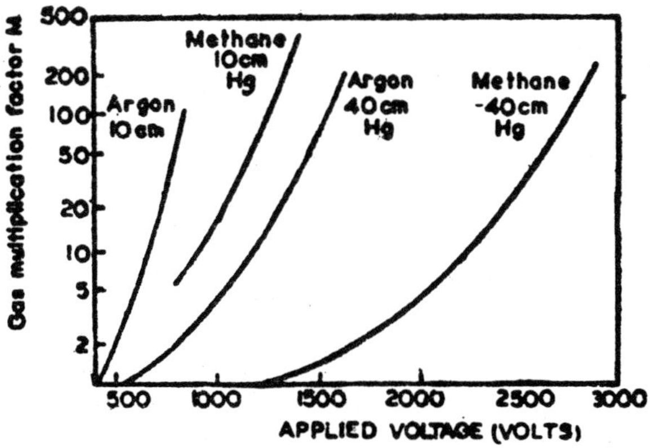

Fig. 2.6 : Variation of pulse size (log) with applied voltage.

Thus, the pulse size remains independent of the potential as long as the voltage is increased upto the point at which secondary electrons arc produced by collision in the gas. This region is known as *ionization cham-ber region or saturation region.*

In region III, voltage is sufficiently high so that electron approaching the central wire attains suffi-cient energy to produce new ions, due to collision with gas molecules. The original electron and the new one, thus, produced, are again accelerated by the field and more ions are still produced by collision. This process is cumulative and an avalanche of electrons is thus produced. The phenomenon fit frequently called the *Townsend avalanche*, in honor of J.S. Townsend who did pioneer work in this subject. The number of the ion pairs formed by collision as each electron travels toward the central wire is known as *multiplication factor M.* The voltage region in which M remains constant is called the *proportional region.* The apparatus operates in this region as a proportional counter. In this pulse height remains proportional to the amount of energy lost in the chamber by the primary ionizing particle.

If the voltage is increased further (*region IV*) pulse heights continue to increase. The heights are still related but no longer proportional to initial ionizing intensity. The voltage region in which M depends some what on the pulse size is called the region of *limited proportionality* and is not much used for measuring. The limitation may be because:

(a) Ultraviolet photons may be formed.

(b) New electrons may be formed when positive ions reach the cathode,

(c) The space charge may distort the electric field.

For still higher voltages the *region V* is reached, in which pulse height is very large and also substantially independent of the initial number of ions. This is called *Geiger region*. The device operated in this region is called a Geiger-Müller counter. In the Geiger region the avalanche associated with the drift of a single electron to one point on the anode initiates avalanches over the entire anode length.

At the end of the Geiger region the counter goes into continuous discharge region,

PROPORTIONAL COUNTER

In the parallel plate ionization chamber if the electrode voltage is increased by beyond the saturation interval up the primary ions produced secondary by collisions with the gas: consequently, the primary ionization is multiplied by a factor that depends on the Geometry of the apparatus and on the applied voltage v the chamber under these condition is ion own as proportional counter.

For a cylindrical geometry, the electric field required to produce secondary ionization is given by

$$E = \frac{V}{r \log_e b / a}$$

Where,

a = radius of central wire

b = inner radius of cathode cylinder

V = electrode or central wire voltage

r = distance of electric field E from the central wire.

The electrons ejected from gas molecules by ionizing particle drift towards the positive electrode. If they move in a region of high electric field E, they will produce additional ionization by collisions with molecules of the gas. In drifting a short distance ax, n electrons will increase the number of elections to n + dn, where

$$dn = \alpha\, n\, dx \qquad ...(1)$$

Here a is the first *Townsend coefficient.* If n_0 is the number of electrons set free by ionizing particle and x is the distance through which these electrons drift to reach the positive electrode, the number of electrons reaching this electrode is given by

$$n = n_0 \exp.\left[\int_0^x \alpha\, dx\right] \qquad ...(2)$$

The ratio n/n_0 is called the *gas amplification or gas multiplication factor M.* The radius of the cylinder and a the radius of an axial wire, the electric field strength at a radius x is

$$E_x = \frac{V_0}{[x \log_e b/a]} \qquad ...(3)$$

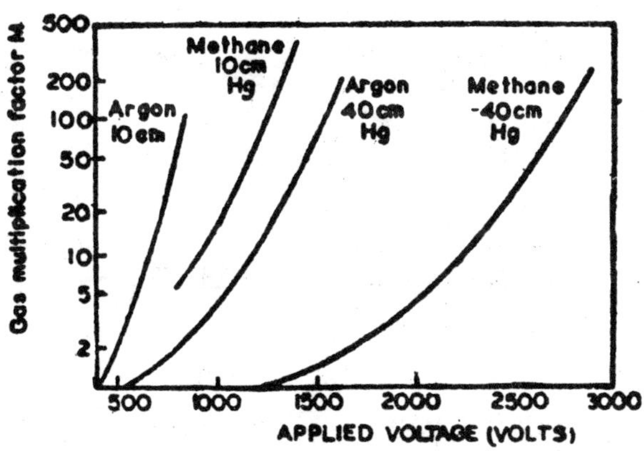

Fig. 2.7 : Variation of M with applied voltages.

Let us assume that every electron produced in the primary ionization gives rise to n-secondary electrons by collisions. The production of secondary electrons during the collision is associated with the photons which in turn produce photoelectrons. If p is the probability of production of photoelectrons, there will be np photoelectrons in all. If each photoelectron produces n-electrons by further collisions, we shall have second-generation avalanche of n^2p electrons, which in turn will give rise to more electrons.

$$\therefore \text{ Total number of electrons } M = n + n^2p + n^3p^2 + ... \quad ...(4)$$

For practical proportional counter $np < 1$, hence $M = n/(1 - np)$. It represents that in a proportional counter the total number of secondary electrons is proportional to the number of initial electrons. The variation of M with voltage in argon and methane at two different pressures is shown in the given Fig. 2.7.

The voltage-time dependence of the pulse is shown in Fig. 2.8. The upper curve shows the behaviour for large values of the time constant. The initial rise of the pulse is due to the high velocity of the positive ions as they recede from the anode. Positive ions then move in a weak field and the pulse approaches its final height extremely slowly.

For the case of a finite time constant the pulse has the shape shown by lower curves. In practice, suitable clipping time constants (RC $\sim 0.1\mu$ sec.) are used in the output circuit so that better time resolutions are available.

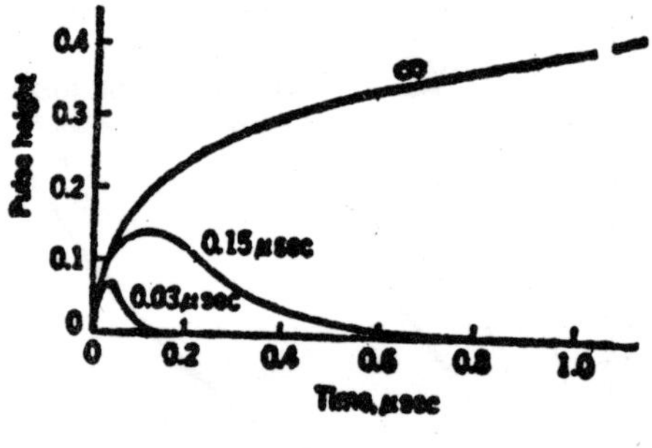

Fig. 2.8 : The pulse space.

If the primary ionizing particles arrive in rapid succession, mechanical counter can not record all the pulses. The output from the amplifier is therefore first passed through a scaling circuit before being fed into a mechanical counter. A sealer is an electronic circuit arrangement that feeds a predetermined fraction of the actual particle counts to the mechanical counter.

CERENKOV COUNTERS

When a charged particle moves through a transparent dielectric medium with a velocity greater than the velocity of light in that medium, weak electro-magnetic radiation is emitted. These radiations are named Cerenkov radiations. The radiation is emitted mainly in the direction of the charged particle and has a continuous spectrum from red to ultra-violet at least.

A charged particle traversing a dielectric medium causes polarization of the molecules adjacent to its path, as the charged particle attracts the oppositely charged component of a molecule and repels the similarly charged component of the molecule. For slowly moving particles, polarized regions are formed in step with the motion in the vicinity of the particle.

The resultant dipole field is zero as the polarization field around the electron is symmetrical in the axial and azimuthal directions, hence no radiation is emitted. For fast moving particles the polarization is *axially asymmetric* but azimuthally symmetric and the resultant dipole field acts even at large distances from the electron's track. The radiation pulses are thus emitted in succession by the molecules along the path of the charged particle. These radiations are like the radiations emitted from a *damped oscillating dipole moment.*

In a dielectric medium of refractive index μ, photons move with a velocity c/μ. Fig. 1.26 (left) shows the three interactions between a charged particle moving with velocity v ($< c/\mu$) and the electrons in the medium. The wavefronts of the electromagnetic waves resulting from interaction do not interrelate and these interactions are entirely independent. Fig. 2.9 (right) shows the interactions when $v > c/\mu$. The wavefronts from 1 and 2 lie inside 3 at the time of the third interaction.

The Huygens construction of elementary wave optics shows that a conical wavefront can be drawn tangent to the spherical surfaces and the Cerenkov radiation can be observed at a particular angle with respect to the direction of motion of the particle.

The particle moves along ABC with velocity v It radiates energy in all directions from the points along its path ABC. The time in which particle travels a distance AC, the radiations will move a distance AE, CE is thus the wavefront. The angle which the Cerenkov Radiation makes with the direction of charged particle is given by

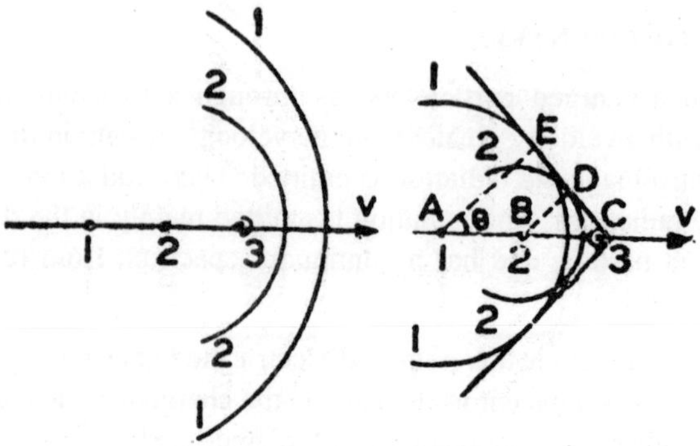

Fig. 2.9 : Wavefronts resulting from radiation losses: with v < c/μ (left) and with v > c/μ (right).

$$\cos\theta = \frac{AB}{AC} = \frac{BD}{BC} = \frac{ct/\mu}{vt} = \frac{c}{\mu v} = \frac{1}{\beta\mu} \quad ...(1)$$

The Cerenkov radiation is not confined to one plane but is propagated along the surface of a cone whose axis coincides with the direction of motion of the charged particle and whose semivertical angle is θ, given by eqn. (33), which shows that:

1. θ depends on the frequency (*i.e.,* wavelength) of the Cerenkov radiation. The radiation arc mainly of the high frequency region of the visible spectrum.
2. Unlike the bremsstrahlung, θ is independent of the mass of moving charged particle.
3. The emission angle increases with increasing the particle velocity β.
4. As P cannot exceed unity the maximum angle of emission.

   $$\theta_{max} = \cos^{-1}(1/\mu)$$

5. For a given constant refractive index μ, there is a threshold velocity β_{th} = l/μ below which no Cerenkov radiation is emitted and at which the radiations are along the particles direction of motion.

Radiation (1) for the constructive interference should satisfy the following conditions:

(a) The particle track length must be large compared with the wavelength λ of the Cerenkov radiation.

(b) The particle must move with the constant velocity in the medium.

Frank and Tamm have shown from classical theory that the total energy radiated in a short element of the particle's path ds is given as

$$\frac{dT}{ds} = \frac{\pi z^2 e^2}{\epsilon_0 c^2} \int \left(1 - \frac{1}{\beta^2 \mu^2}\right) v\, dv$$

where ze is the charge on the moving charged particles, v is the frequency of the emitted radiation, and the integral extends overall frequencies for which $\beta\mu > 1$. This loss is quite small (*e.g.*, about 10^3eV/cm for singly charged particles moving through glass or lucite) and is negligible compared to the ionization losses.

Consider the Cerenkov radiation emitted between two frequencies v_1 and v_2 as composed of quanta whose average energy is $hv = 1/2h(v_1 + v_2)$, then from eqn. (44) the average number of quanta emitted per unit length is

$$\frac{1}{h_v} = \frac{dT}{ds} = \frac{\pi z^2 e^2}{\epsilon_0 hc^2}(v_2 - v_1)\left(1 - \frac{1}{\beta^2 \mu^2}\right) \quad ...(2)$$

where μ if the average refractive index of the medium over the frequency interval v_1 to v_2.

Relation (2) shows that for extreme relativistic electrons passing through glass ($Z = -1$, $\beta \simeq 1$, $\mu \simeq 1.5$) the (Cerenkov angle is 48° and the Cerenkov photons emitted per cm within the visible spectrum ($\Delta v = v_2 - v_1 \simeq 3 \times 10^{14}$ sec^{-1}) are 200. As the number of quanta per unit path length varies as I/λ^2, the short wavelength are thus predominantly present in Cerenkov radiation.

The properties discussed above have been extensively utilized in Cerenkov detectors which can be used not only to detect swiftly moving charged particles but also to determine their velocity. These detectors possess the following advantages:

1. Dependence of intensity on particle velocity.
2. Directional emission of light, with an angle dependent on velocity.
3. High efficiency and high counting rate.
4. The duration of the light pulse at any point with in the Cerenkov cone is vanishingly short ($\sim \leq 50^{-10}$ sec).

R.L. Mather has observed Cerenkov radiations radiated by a collimated beam of 340 MeV protons passing through an optically flat sheet of dense flint glass of 0.67 mm thickness with a 35 mm Leica camera. The glass sheet, whose rear surface was alumunized, was set with respect to the proton beam so that part of the Cerenkov cone radiated in the glass would incident normally on the aluminium surface and reflect in such a way that there would be no refraction at the front surface of the glass.

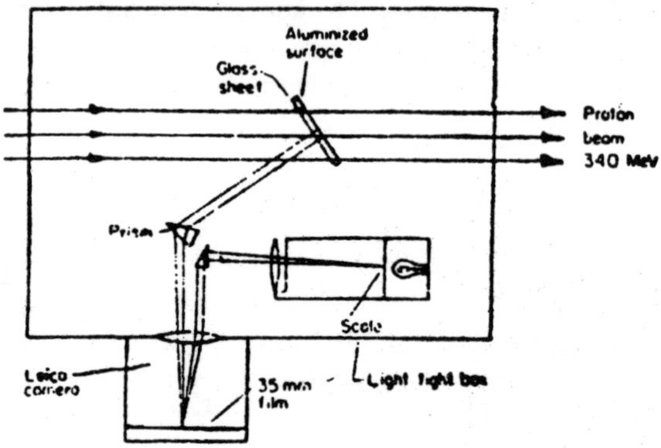

Fig. 2.10 : Measurement of velocity of high energy protons.

The angle of emission of radiations varies with wavelength λ because the refractive index depends on λ. This dispersion effect was decreased to nearly zero by Mather after introducing a prism. The recorded Cerenkov radiation on the camera film corresponded to a θ value of 38.5°. Using the relation (2) β comes out to be 0.680. The kinetic energy E_{kin} of the protons can be obtained as

$$E_{kin}\ m_p c^2 (1 - \beta^2)^{-1} - m_p c^2 + 341 \text{ MeV} \qquad ...(3)$$

This value was found to have 1% accuracy.

J. Marshall designed a good detector, in which the Cerenkov light was focused on to the cathode of two photomuitiplier tubes for recording high energy π-mesons (pions). A collimated beam of pions was allowed to pass along the axis of a large hemispherical perspex lens. The radiation radiated in this lens was reflected by a surrounding cylindrical mirror M_1 on to one of the plane mirrors M_2, M_3 and then on to the cathodes of the photomuitiplier tubes (lP 28), which were operated in coincidence. The position of the radiator lens was adjusted so that the Cerenkov light cone was correctly focused on to the photo tubes to give maximum coincidence counting rate. For 145 MeV pions the energy resolution is 13%.

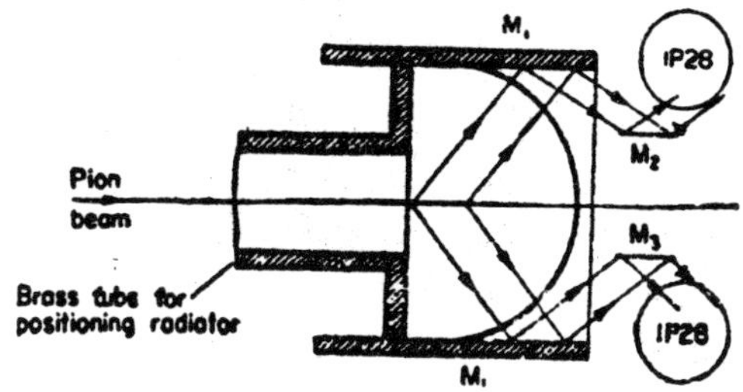

Fig. 2.11 : Marshall's focusing Cerenkov detector.

Cerenkov counters have been designed for charged particles in a narrow velocity range. Cerenkov detectors in which the light is radiated by a gas have been used for highly relativistic charged particles. Because the radiation in a given medium is produced only when the speed of the charged particles is greater than the velocity of light in that medium, this counter can be used as a *threshold detector, i.e.*, to detect only particles whose speed exceeds a certain value.

These counters are essentially *directional counters* became 'the Cerenkov radiations are emitted in a forward cone of an angle given by eqn. (3). The direction of the incoming particle can, therefore, be measured. The angle θ depends on the velocity v of the incident particle. This counter can thus be used to measure the velocity of incoming particle. Particles of *different masses* can be distinguished in two ways: first, for the same velocity, *i.e.,* the same Cerenkov angle, the brightness of the flash is proportional to the square of the atomic number and, second, for the same K.E. or range, the velocity of the particle is inversely related to the square root of the mass number. The particles of different charge can also be distinguished by these counters.

The blue glow seen around nuclear reactors of the swimming pool type reactor is almost due to *Cerenkov radiation*, produced by high speed electrons (Compton electrons produced in γ-ray scattering events). Since the core or fuel elements are in extended source and the emitted electrons show a distribution over a wide range of energies and angles. The radiation thus appears diffuse. The energy and momentum of charged particles (electrons) can thus be evaluated.

CLOUD CHAMBER

During the stay in the Scottish hills, in 1984, he started experimenting under laboratory conditions and intended to study the optical interesting phenomena appearing from illuminated fogs. After two years he discovered the fundamental principle of the cloud chamber and designed the first cloud chamber based on the principle in 1911.

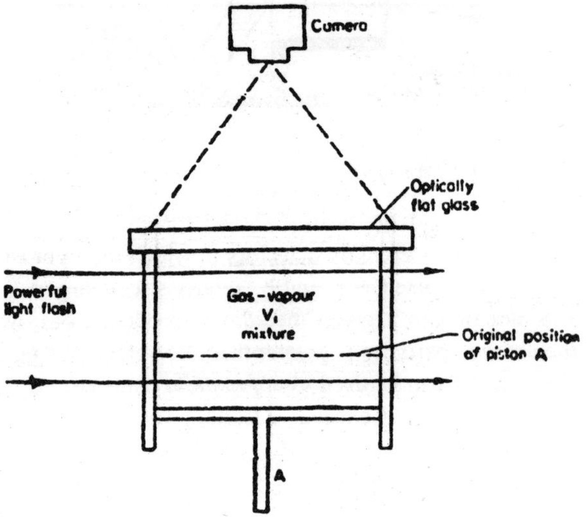

Fig. 2.12 : Schematic diagram of Wilson expansion cloud chamber.

Air mixed with saturated vapour of water or any other liquid, *e.g.*, alcohol or ether contained in a cylinder fitted with a piston can be expanded rapidly by a fast but relatively short motion of a piston. This adiabatic expansion will result in fall of temperature and super-saturation of vapour. The vapour will condense in cloud of water droplets.

The production of such a cloud is however, impossible unless nuclei, on which the water vapour may condense are provided. If particles of dust are present in the air (as an ordinary cases) they will act as condensation nuclei. But in dust free air this phenomenon of condensation and formation of cloud will, therefore, be extremely difficult. C.T.R. Wilson discovered that the condensation cloud could be produced even in dust free air provided the air was ionized by an ionizing agent such as the X-rays, cathode rays or any nuclear radiations.

If ionizing particle entered the chamber either immediately before or immediately after the expansion, the ions left in its path would act as condensation nuclei. This experiment shows that as super saturation increases, the negative ions first serve as centres of condensation, then as the volume increases both positive and negative ions serve as the nuclei of droplets. A close array of fine droplets, *i.e.*, a kind of linear cloud, called a *cloud track*, will thus be formed. By using suitable strong illumination from the side, the track appears as a white line on a dark background. This can be photographed by means of a camera so that a record can be obtained.

Let V_1 be the volume, P_1 the saturated vapour pressure and T_1 the temperature of the vapour. Its mass m_0 (in moles) is determined by the relation

$$P_1V_1 = m_0RT_1 \qquad ...(1)$$

If the vapour is expended adiabatically from V_1 to V_2. The new temperature T_2 is given by

$$T_1V_1\gamma^{-1} = T_2V_2\gamma^{-1} \qquad ...(2)$$

If no vapour condenses, then the new pressure is given by

$$P_2V_2 = m_0RT_2 \qquad ...(3)$$

The vapour is in supersaturated condition. If ions are present in the space, the droplets will condense on them till the pressure falls to P_2. If m is the mass of the vapour remaining uncondensed, then

$$P_2V_2 = mRT_2 \qquad ...(4)$$

The degree of supersaturation, is the main determinant governing the growth of droplets about ions and may be defined as the ratio of the actual density of vapour to the saturation density, Thus, we have

$$S = \frac{m_0}{m} = \frac{P_2}{P_1} = \frac{P_1}{P_2}\left(\frac{V_1}{V_2}\right)^{\gamma} = \frac{P_1}{P_2} \times \frac{1}{(\text{Expansion ratio})^{\gamma}} \qquad ...(5)$$

The pressure of vapour in equilibrium with small drops of liquid is not the same as that for a plane liquid surface. For a drop of radius r and charge ze the saturated vapour pressure P_r is given by

$$\log \frac{P_r}{P_\infty} = \frac{M}{\rho TR}\left[\frac{2\theta}{r} - \frac{z^2e^2(k-1)}{32\,\epsilon_0\,\pi^2 k^{\gamma 4}}\right] \qquad ...(6)$$

The formula for ze = 0 was derived by Lord Kelvin and the electrostatic term was later introduced by J. J. Thomson. In this relation P_∞ a and M are respectively the normal saturated vapour pressure for plane surface and the molecular weight of the vapour and ρ, ∞ and R are respectively the density, surface tension and dielectric constant of the liquid.

The Wilson cloud chamber makes it possible to study the behaviour of individual atoms, to photograph the actual path of ionizing radiations and to analyse at a spare time the complicated interactions which may take place between charged particles and individual atoms. Since the year 1911 the apparatus has been improved in many ways, although the fundamental principle remains unchanged.

P.M.S. Blackett and G.P.S. Occhialini invented the Geiger counter controlled expansion cloud chamber for the study of cosmic rays. Such chambers have been used extensively in T.I.F.R. Bombay and Bose Institute Calcutta, in India, for cosmic ray work. One such design is shown in Fig. 2.13.

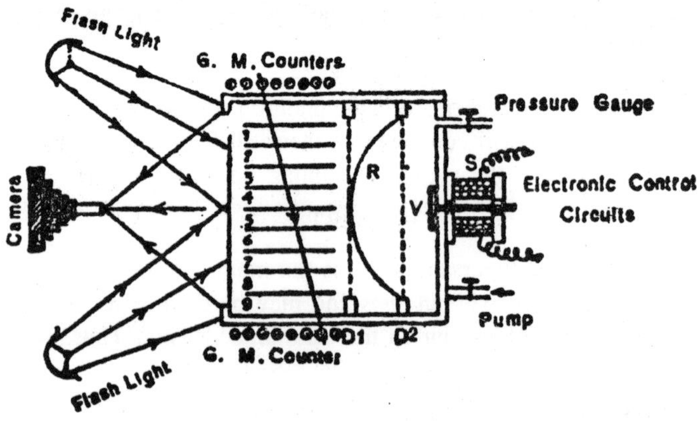

Fig. 2.13 : Schematic diagram of automatic.

In this arrangement two sets of G.M. counter trays are placed one at the top and other below the bottom of the active volume of the chamber. These trays are connected in coincidence so that any event due to the passage of a particle is triggered by these coincidence counters. The chamber is kept at a higher pressure with the help of compression pump which presses the membrane of rubber R. R is kept under tension and rests on the perforated metal disc D_1. R magnetic valve V, electrically controlled with the help of a solenoid S, is kept in position by energising the solenoid.

When the counters (in coincidence) register an event, the electric pulse produces an impulse of current. The impulse is given to the solenoid in opposite direction. The valve V is now released and the gas escapes out bringing the membrane of rubber R to the position of the perforated disc D_2. After expansion, light is automatically turned to illuminate the tracks, the chamber volume is photographed. The ions are swept from the chamber by application of a clearing field and both chamber and camera are reset to await the next event.

In Wilson Cloud Chamber water is used to saturate the air but it becomes more common to employ ethyl or propyl alcohol or a mixture of alcohol and water. The use of alcohol gives better condensation on positive ions and the extent of expansion necessary for droplet formation is diminished from 1.25 to about 1.10 at ordinary pressure. Although, air is the usual gas, cloud chambers containing argon are sometimes employed. The higher pressures are desirable for the study of high energy particles. In cosmic-ray work dense metal plates have often been placed across the cloud chamber to ensure interactions of the high energy particle.

The heavy, slow particles like α-particles produce broad, densely packed, straight line tracks. Near the end of their path the particles may suffer sharp deflection as the result of impacts with the nuclei of oxygen or nitrogen present in the air. Slow electrons produce narrow, beaded, tortuous tracks, while fast particles both light and heavy produce narrow, beaded, straight tracks. By counting the drops in the cloud track the specific ionization can be determined and the nature of the particle identified. The sign of the electric charge and the momentum p of the particle can be determined if the chamber is placed in a strong magnetic field. If a stream of a particles, each carrying a charge q, moves initially in a straight line with velocity v and a magnetic field B is applied in a direction at right angles to the direction of motion, the panicles will be forced by the field to follow a circular path of radius R. The magnetic force Bqv is exactly balanced by the centrifugal force mv_2/R. Thus

$$Bqv = mv^2/R \text{ or } p(= mv) = BRq \qquad ...(7)$$

The kinetic energy E of the particle can be easily calculated, if the rest mass energy m_0c_2 of the particle is known, by the relation

$$E = \sqrt{[p^2c^2 + (m_0c^2)^2]} - m_0c^2 \qquad ...(8)$$

The *disadvantage of the cloud chamber* lies in the fact that it needs a definite time to recover after an expansion and hence it is not possible to have a continuous record of events taking place in the chamber. A continuously sensitive, known as *diffusion cloud chamber* has been developed by A. Langsdorf in the United States in 1939. It consists of a vessel containing air or other gas, kept warm at the top and cold at the bottom.

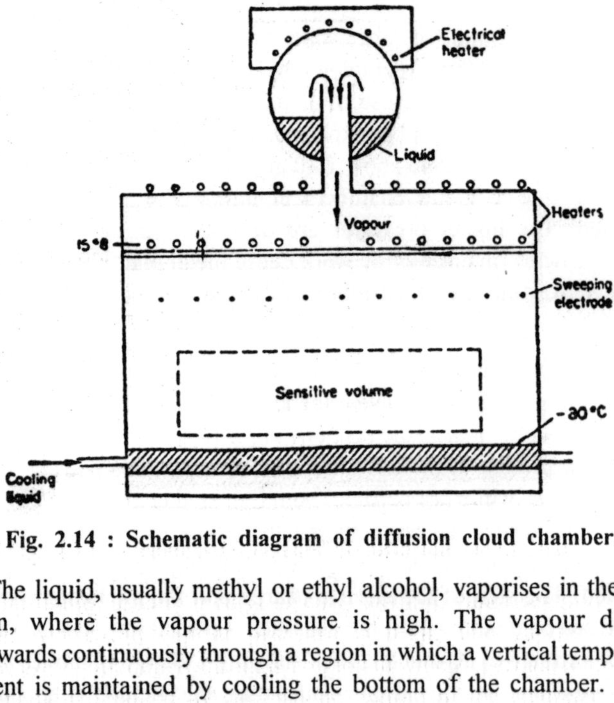

Fig. 2.14 : Schematic diagram of diffusion cloud chamber.

The liquid, usually methyl or ethyl alcohol, vaporises in the warm region, where the vapour pressure is high. The vapour diffuses downwards continuously through a region in which a vertical temperature gradient is maintained by cooling the bottom of the chamber. As the vapour diffuses into the colder regions of the gas, the saturation vapour pressure decreases rapidly and there is a volume near the cold base of the chamber where the supersaturated condition is maintained and thus eliminating the insensitive cycling time of the conventional cloud chamber.

If some ions are present in this region, vapour forms droplets around them. In order to increase the frequency of occurrence of nuclear events, the pressure of the gas in the chamber is increased. *The main drawback of the diffusion chamber is that the sensitive region is no more than 7.5 cm deep.*

The diffusion cloud chamber is a continuously sensitive chamber, has many uses in the study of high energy particles obtained from accelerators. Its interesting application is the determination of the half-life of the free neutron. The discovery of the positron by electron positron pair production method, K^o meson by observing the decay in flight into pions, hyperon Λ^o by the observation of the decay in flight and that of muons in secondary cosmic radiation were made with the cloud chamber. The first reasonably accurate determination of the rest mass of the muon was made by W, B. Fretter with the help of cloud chamber.

SPARK CHAMBER

The spark chamber utilizes an incipient electrical discharge in a gas, like that used in a Geiger counter, to give track geometry information, like that provided by a bubble chamber. This has found increasing use in the field of high energy physics. The basic principle of operation is as follows:

A high voltage is maintained between two plates spaced in a vessel containing a gas but the electric field is not quite enough to permit the passage of a spark. Now if an ionizing particle enters the gas space, a spark will pass and it will tend to follow the path of the ion pairs produced by the ionizing particle. The passage of a spark through a gas that had been ionized by nuclear radiation was used by H. Greinacher, in 1935, for the detection of α-particles. F. Bella and C. Franzinetti in 1953, in Italy, confirmed the localization of the spark discharge and the first photographs were given in 1955 by P.G. Henning in Germany. The modern spark chamber stems from the work described by the British physicists T. E. Cranshaw and J. F. de Beer in 1957. In 1959, S. Fukui and S. Miyamoto in Japan found that with neon several particle tracks could be observed simultaneously in the spark chamber and with air only one particle track could be formed at a time. Spark chambers were used by Sree Kantan and coworkers at T. I. F. R. Bombay in 1960, to trigger high energy cosmic-ray events.

In its simplest form spark chamber consists of a series of large thin parallel metal plates (aluminium), several square feet in area, spaced about 1 cm. apart in a chamber filled with a neon gas at atmospheric pressure. All the plates are isolated from each other. The first, third, fifth, etc., plates are grounded and the second, fourth, sixth, etc., plates are connected to a high voltage direct current pulse generator, which gives them a high potential in short bursts of the order of a microsecond each. This potential is just enough to cause sparks to occur between the plates in such regions as are ionized by a particle entering the chamber. The chamber is sensitive for about half microsecond preceding the application of the voltage.

Any charged particle passing through the chamber during this period produces along its path a visible and audible spark discharge, which can be located by photographing or by other methods. The photograph of all the individual spark reveals the sections of the trajectory of the particle between the plates. Ordinarily two cameras at right angles to one another are used to photograph.

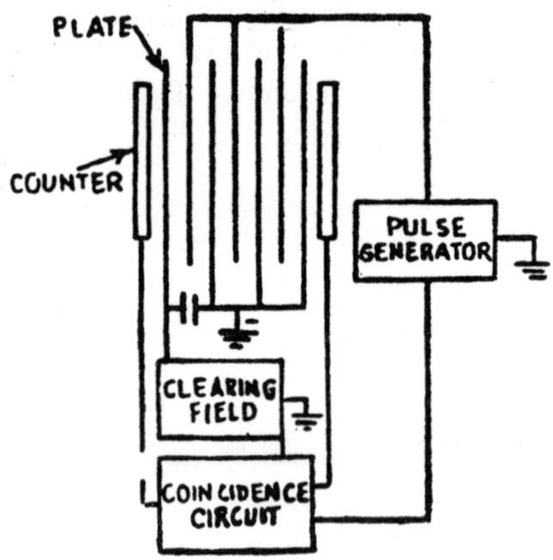

Fig. 2.15 : Schematic representation of spark chamber circuit.

In order to prepare the chamber for recording the path of another particle, any ion-pairs present between the plates must be removed, by applying a clearing field of about 200 volts per cm. continuously in the direction opposite to that of the high voltage spark field. Due to this field a spark can form only if the high voltage is attained within about a microsecond after the entry of an ionizing particle. In practice this delay time is about 0.2 microsecond. The recovery time depends on the gas, the design of the chamber and the energy of the spark. It is usually between one to 20 milliseconds, which is still very short in comparison with the recovery times of cloud and bubble chambers.

The chamber can be operated in a magnetic field and the momentum and charge sign of the ionizing particle can then be determined from the curvature of the track. The number, thickness, spacing and material of the plates in a spark chamber are varied to suit the experimental requirements. The plates are commonly made of very thin aluminium to minimize energy absorption from the particle and are closely spaced to define the track more accurately. The sparks tend to be perpendicular to the plates in their passage between the adjacent plates. The photograph of spark chamber tracks and their interpretation. Sometimes it is desired to study the interactions of specific particles with matter. The plates are then thicker usually and are made of either carbon or a metal, such as iron, lead, tantalum or brass. The interactions take place almost within the solid plates as the density of nuclei in the gas between the plates is small.

One of the main advantages of the spark chamber over the other chambers is that the triggering and removal of ions by the clearing field are comparatively simple. With the help of this chamber it is possible to study an individual event produced by one particle among the many passing through the chamber at a rate that may be as high as a million per second. The spark chamber event of interest is selected by an arrangement of counters, called a *Logic circuit.*

Analysis of spark chamber photographs can be carried out by ordinary calculations or by computer techniques. The preliminary scanning stage may not be required because spark chamber events of interest have been selected by the logic system. The tracking, geometry and kinematic analysis arc performed in the manner as with bubble chamber pictures.

There are *two main drawbacks* to a spark chamber as compared with a bubble chamber. First, in the bubble chamber the liquid may be chosen so as to restrict the interactions of the incoming particle to those of a special type. Second, the co-ordinates of tracks and of the points of interaction in a bubble chamber can be defined with much greater accuracy than is possible in a spark chamber. As a general rule, bubble chambers are preferred where good space resolution is desirable. On the other hand, spark chambers permit the selection of events of interest with a high repetition rate, *i.e., they have better time resolution.*

Since 1963, various "filmless" spark chambers have been developed in which photography of the spark discharge is avoided. In such devices, the co-ordinates of the spark tracks are obtained directly in a form suitable for a computer to determine the geometry of the event. The grounded plate of every pair is replaced by an array of closely spaced parallel wires. For the high voltage plates a similar arrangement of wires could be used but solid plates are generally employed since they are more convenient and easier to fabricate. The wires arc made from a magnetostrictive material (50% Fe and 50% Co alloy) in *one type of wire spark chamber*. A passage of a spark from a particular wire produces a local magnetic field and a deformation occurs at the affected point.

The deformation travels then along the wire to suitable detectors at each end. The ratio of the times between the firing of the spark and arrival of the deformation at the ends gives one co-ordinate, the other co-ordinate of the spark track is obtained by identifying the particular wire at which signals were received. The most promising wire spark chamber, referred to as a *spark-chamber hodoscope*, differs from the acoustic wire chamber in respect that the wires constituting the alternate grounded plates are alright angles to each other. At one end each wire is threaded through a ferrite core, such as is used in computer memory devices.

A ferrite material has two stable states of magnetization and by a magnetic field (from an electric current) either one can be converted to the other. The magnetism of ferrite cores associated with the wires from which the spark passes is changed from one state to the other by the spark current or we can say that ferrite cores are nipped. By interrogating the cores in a manner similar to that employed in a computer the locations of the affected wires are determined and flipped cores are reset at the same time. Since successive planes of wires are at right angles, the information from the cores can give the co-ordinates of the spark track in three dimensions in digital form.

It has a advantage of a very short recovery time. The amount of energy required to flip the core is very small and a feeble spark, which would not give off enough light to be photographed, is adequate for this purpose.

Sometimes it is desirable to make observations of interactions at very-very short intervals with high energy particles. *The scintillation counter hodoscope* can be used for this purpose. Although the accuracy of determining the co-ordinates of particle tracks decreases. It consists of an array of long thin rectangular strips (25 × 0.5 × 0.5 cm) of a plastic scintillator material. These strips are set side by side to produce a sensitive planer area. A photoelectric tube is associated with each strip and the passage of a charged particle through a given strip is indicated by the output signal from the corresponding photo-tube. This provides one co-ordinate of the particle track. For the determination of co-ordinates in the other two directions, it is necessary to build a hodoscope from several layers of scintillators arranged with the strips in successive layers at right angles.

The output signals can be fed directly to an on-line computer for analysis. This type of hodoscope can record events at a much faster rate than wire spark chamber hodoscope can. It is because of the very short recovery time of a scintillation counter. The accuracy with which the track co-ordinates can be determined in a scintillation counter hodoscope is limited as it is dependent on the width of the scintillator strip.

CHAMBER OF IONIZATION

If we applied a small potential difference to the electrodes, the –ve ions (or electrons) will be attracted to the anode and the +ve ions will go to the cathode. At the surface of the anode there will be a high concentration of negative ions. Recombination in this region will be a small because the rate of recombination is proportional to the product n^+n^- and here n^+ is very small. All positive ions produced in this region move out promptly towards the cathode. Therefore, the recombination will be maximum at the center of the space.

A small current will flow through a resistor connecting the electrodes externally As the voltage between the electrodes is increased, the ions will move faster, the number of recombination will be minimum and the pulse size will thus be increased. The pulse size increases with the voltage, until it reaches to a value corresponding to the no recombination of ions. This value of current is called the *saturation current* and the chamber is said to reach in *saturation.* An ionization chamber, in its simplest form consist of vessel usually field with gas and into which it inserted an insulated electrodes. The gas becomes ionized when charged particles pass through it. If no voltage is applied to the container there will be no directing force acting the ions and no current will flow through the gas.

The ionisation chamber works in the region of constant pulse size. In this region the applied voltage is high enough to prevent recombination of ions and low enough to prevent multiplication.

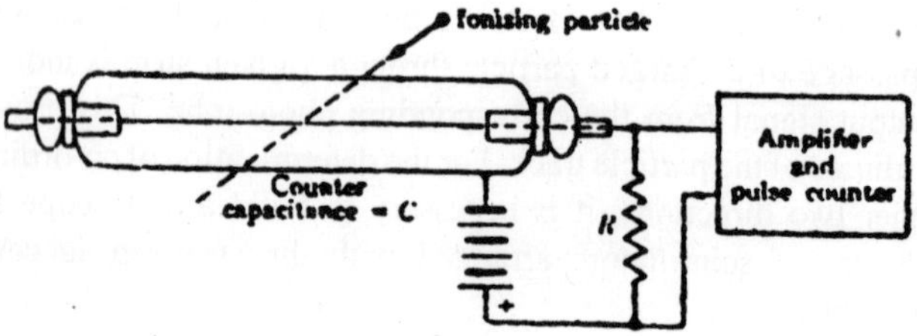

Fig. 2.16 : Ionization chamber.

In the initial stages, the gold-leaf electroscopes of great sensitivities or the quadrant electrometers were used. The next step in the development was a modification of the electroscope, designed by Lauritsen, where the gold leaf was replaced by a tiny metal coated quartz fibre. It could not give quick response to particles that follow one another in rapid succession.

This difficulty was overcome to a great extent by a device known as linear or dc amplifier, which increases, without distorting, the feeble ionisation current from a single particle to a value which can easily register on an oscillograph or an electrical counter, dc amplifiers are more susceptible to disturbance and drift and more difficult to arrange with several successive stages of amplification than those designed for amplifying alternating currents.

Ionization chamber instruments may be grouped into two categories, depending upon the value of the time constant RC of the system relative to the frequency of arrival of the ionizing events. If RC is long compared to time between ionizing events a steady state is reached and a *direct current* may be measured. On the other hand, if RC is small each pulse produced by an ionizing event may be detected separately. These two types are also known as *integrating and non-integrating* respectively and the instrument as current and pulse ionization chambers.

Let a track of n ion pairs is formed parallel to and at a distance x_0 from the axial wire The electrodes are parallel and separated at a distance d. A voltage V_0 is applied to the electrodes, so that electrons and positive ions drift in opposite directions in a uniform electric field V_0/d. Let us first assume that the time constant RC (where C = capacitance of the chamber + input capacitance of the amplifier) is much greater than the collection time of ions. So that the current through R may begin as soon as the electrons and +ve ions move apart in the chamber.

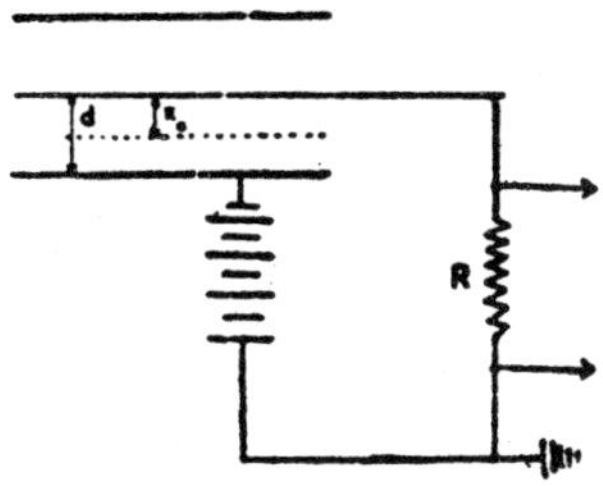

Fig. 2.17 : Ionization in an ionization chamber.

The change of potential at the positive electrode at the time t, due to a track of n ion pairs formed at t = 0 is given as

$$V(t) = n\,[q_e(t) + q_p(t)]/C, \qquad ...(1)$$

where $q_e(t)$, the charge induced on the positive electrode at time t by a single electron, and is given as

$$q_e(t) = -e.\frac{\text{P.D. between the electron at time t and the -ve electrode}}{\text{P.D. between the electrodes}}$$

$$= -e.\frac{d - x_e(t)}{d} = -e\frac{[d - (x_0 - v_e t)]}{d} \qquad ...(2)$$

and $q_p(t)$, the charge induced on the positive electrode at time t by a single positive ion, is given as

$$q_p(t) = +e.\frac{\text{P.D. between the +ve ion at time t and the -ve electode}}{\text{P.D. between the electrodes}}$$

$$= e.\frac{d - x_p(t)}{d} = e\frac{[d - (x_0 - v_p t)]}{d} \qquad ...(3)$$

The relations (2) and (3) are only true for parallel plate geometry in which potential gradient is uniform.

From eq. (2), (3) and (1) we get.

$$V(t) = -ne(v_e t) + v_p t)/Cd. \qquad ...(4)$$

As the electron drift velocities v_e are about a thousand times greater than the heavy ion velocities v_p, hence, for the time between 0 and t_e is the electron collection time ($x_0 = v_e t_e$), the above relation transform to

$$V(t) = -nev_e t/Cd. \qquad ....(5)$$

For $t > t_e$, there is an abrupt decrease in the rate of rise of potential, the new rate being v_p/v_e times the original rate. The change of potential becomes—ne/C when $t = t_p$, the time in which positive ions reach the negative electrode.

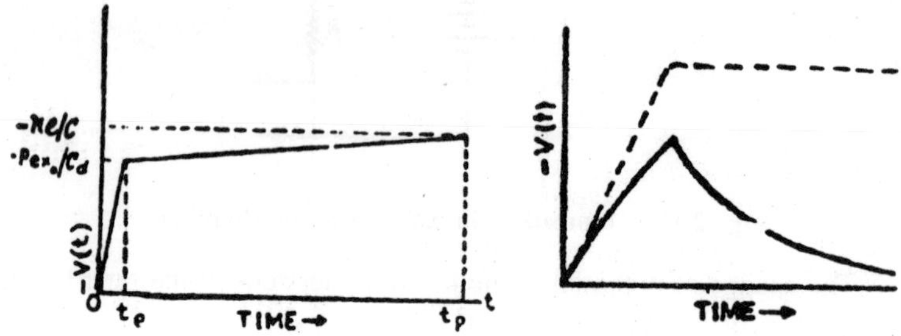

Fig. 2.18 : (a) Shape of potential pulse. (b) Pulse cliping.

This limit is proportional to n and is independent of where the ions were formed in the chamber. The shape of the collector potential curve does depend upon where the ions are formed.

With sufficient amplification, a large pulse will appear at the amplifier output terminal. To have the height of the output pulse proportional to the amount of ionization produced by the particle in the chamber, two kinds of proportional or linear amplifiers can be used. One, known as slow *amplifiers*, in which the shortest time constant is chosen long compared with the drift time of the +ve ions, have several disadvantages. The others, known as fast amplifiers, in which time constant RC is made short enough so that the induced potentials by the +ve ions are not of interest and clip the pulses.

Thus the slow linear rise, after the collection of electrons, is removed and the height of the recorded pulses depends on $V(t_e)$ and not on $V(t_p)$. If the complete range of an *ionizing particle is confined to the region between the negative electrode and grid*, the grid screens the +ve electrode from the electrons and +ve ions. Electrons, while drifting through the grid, induce a change of potential at the +ve electrode. As positive ions are screened from the +ve electrode by the grid, hence have no effect. Thus, the pulse height increases due to the induction effect of the moving electrons. When all the electrons are collected it attains a maximum value given by

$$V_{max} = -\frac{ne}{C} \qquad \text{...(6)}$$

It is also independent of the position and orientation of the ionization track.

The height of the amplified pulses from the ionization chamber

$$V = -\frac{AEe}{WC} \qquad ...(7)$$

where A is the voltage gain of the amplifier, E the energy of the ionizing particle, e the electronic charge and W the mean energy required to produce an ion pair in the gas filled in the chamber.

BUBBLE CHAMBER

To detect very energetic particles the Cloud Chamber is not very suitable since the interaction of energetic particles cannot be completely observed in the chamber. To overcome this disadvantage it is necessary to use a chamber filled with a substance having a much greater stopping power. This has led to the development of the Bubble Chamber. In 1952, D.A. Glaser, at the University of Michigan, designed and made the first bubble chamber and this has now become one of the largest and at the same time one of the costiliest type of particle detector in use.

We know that normally the liquid boils with the evolution of bubbles of vapour at the boiling point. If the liquid is heated under a high pressure to a temperature well above its normal boiling point, a sudden release of pressure will leave the liquid in a super heated state. If an ionizing particle traverses the liquid with in a few milliseconds after the pressure is released, the ions left in the track of a particle act as condensation centres for the formation of vapour bubbles. The vapour bubbles grow at a rapid rate and attain a visible size in a time of the order of 10 to 100μ sec. Upon this principle chamber operates. The bubbles are like the droplets in a cloud chamber, visible under strong illumination. If nuclear reactions take place in the liquid of the bubble chamber, the sets of tracks diverge from the collision centre as in the cloud chamber. The tracks are photographed against a dark background.

The time sequence of events for a typical chamber is shown in Fig. 2.19. At the instant A, the high pressure P_1 (400 units) applied to the bubble chamber liquid is suddenly reduced to the lower pressure P_1 (100 units). The pressure starts to rise due to the bubble formation. The liquid is recompressed at C and the chamber returns to its initial position at A'. The time interval between A and A' gives the cycling rate of the chamber. The expansion is timed so that the particle enters the chamber at X and the bubbles are illuminated and photographed at the moment Y.

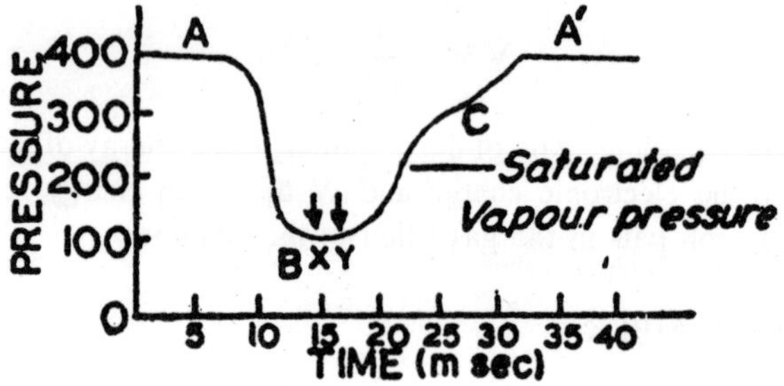

Fig. 2.19 : Time sequence of events in bubble chamber.

The liquids suitable as the working substance of a chamber are transparent and having a low surface tension and a high vapour pressure. The medium in first bubble chamber was ordinary ether, *i.e.,* diethyl ether, but liquid hydrogen, deuterium, helium, propane, xenon and other liquids have been utilized successfully for different purposes. In some experiments it is desired *to study the interaction of particles with protons*, the liquid in the chamber is then hydrogen.

Propane (C_3H_8) is commonly used in the bubble chamber to avoid the need for low temperature (–246°C) operation with liquid hydrogen. It has almost 1.3 times as many hydrogen nuclei (*i.e.,* protons) per unit volume, but the carbon nuclei present are sometimes a drawback. For *observations on neutron interactions* liquid deuterium with a neutron fairly loosely bound to a proton if employed.

Liquid helium chambers operating at 4°K and 1 atm. have been used for studying high energy interactions with the nucleons in the He^4 nucleus. A liquid containing heavy atoms is favoured where stopping power is the main objective, *e.g., in studying the interactions of neutral particles and gamma rays.* Examples are the freon type compound, trifluoro-bromomethane (CF_3Br) and liquid xenon. The former can be used just above room temperature (30°C, 18 atm.) however there are three distinct complex nuclei present in this liquid. Liquid xenon operating at 14°C is ideal for studying interactions with high energy γ-quanta and neutral pions. For satisfactory bubble tracks a small amount of ethylene is to be added. The operating conditions for bubble chambers with different fluids are given in the following Table 5.

Table 5 : Operating conditions for Bubble Chamber fluids

Fluid	Temp. °C	Pressure (atmospheres)	Mean free path for 100 MeV γ-rays (m)	Density (Kg m^{-3})
Hydrogen	–246	5	27	60
Deuterium	–241	7	20	130
Helium	–269	1	18	130
Propane	58	21	2.2	430
Pentane	157	23	—	500
Xenon	–20	26	0.07	2300

Liquid hydrogen is an ideal substance for studying high energy interactions with protons, as it is a pure proton target having very small Coulomb scattering. L W. Alvarez at the Lawrence Radiation Laboratory, California University, thought that if the chamber was sufficiently large and the pressure of the compressed liquid was decreased rapidly, good tracks could be obtained in the interior in spite of the fact that bubbles were formed at the walls.

This conjecture proved to be correct and a chamber six feet long and of cross-section 1½ feet, containing 520 litres of liquid hydrogen, was first operated in 1959 successfully. A schematic diagram of a hydrogen bubble chamber, operating at a temperature of 27°K and developed by Alvarez and his co-workers is illustrated in Fig. 2.20. The vessel was made of stainless steel with glass ports at the top for the viewing cameras. A box of thick glass walls was filled with liquid hydrogen and connected to the expansion pressure system. To maintain the chamber at constant temperature it was surrounded by liquid nitrogen and liquid hydrogen shields.

From the side window W the high energy particles were allowed to enter the chamber. Liquid hydrogen was kept under pressure first but when a particle has passed through it the pressure was released so that it was super heated at a lower pressure. A sudden release of pressure from the expansion valve we followed by light flash and camera took the steroscopic view of the chamber.

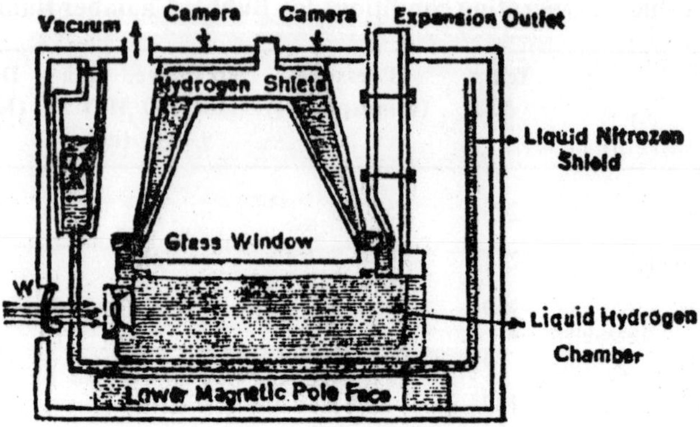

Fig. 2.20 : Schematic diagram of Alvarez bubble chamber.

This was the first of the so-called dirty bubble chambers. Subsequently, 80 inch hydrogen bubble chamber was built at the Brookhaven National Laboratory and similar one is in use at CERN (*Conseil Europeen Pourala Recherche Nucleaire*). Even larger bubble chambers, upto 14 feet in length, are being designed. Similar to the original cloud chamber bubble chamber is not continuously sensitive.

(i) A dirty bubble chamber is sensitive for only a few milliseconds after the pressure reaches its minimum.

(ii) The relatively slow expansion of the liquid does not permit the process to be initiated by an ionizing particle continuously.

(iii) The bubbles formed spontaneously at discontinuities and they interfere with the photography of the particle tracks.

(iv) Once the liquid has been expanded, it takes some time to recover sufficiently for another set of tracks to be recorded.

These drawbacks are overcome by using bubble chambers in connection with high energy particle accelerators, which function in a pulsed manner. The chamber can be made sensitive each time a pulse of particles emerges from the accelerator, and the large density and large volume of the chamber permit appreciable numbers of interesting collision processes even when the incident particles have energies of thousands of MeV. The timing is such that the particle beam from the accelerator enters the bubble chamber just prior to the pressure minimum. A few milliseconds later, powerful lights flash on and the tracks are recorded on a photographic film which moves forward automatically after each exposure. For accelerators, which emit particle pulses at more frequent intervals, smaller bubble chambers, with shorter recovery times, are employed.

A magnetic field is usually provided at the chamber in order to distinguish the sign of the charged particles and to measure their momenta from the radius of curvature of the bubble track. For particles of high momentum, the curvature is small and hence is difficult to measure unless a strong magnetic field is used. Thus, to analyse the collisions it is necessary to maintain the whole chamber in a strong uniform magnetic field (~20,000 gauss). To achieve this a magnet weighing several hundred tons is required with drastic cooling arrangement. Consequently, there has been much interest in the possibility of constructing superconducting magnets for use of bubble chambers.

The problem in the construction of such superconducting magnets of appreciable size has been the difficulty of producing cable with satisfactory physical and mechanical properties at a reasonable cost. In recent years, certain metallic compounds (alloys) of niobium, *e.g.*, with tin, titanium, or zirconium, have been developed for fabrication into super conducting cables. The first successful operation of a bubble chamber, 10 inches in diameter, with a super conducting magnet was achieved at the Argonne National Laboratory in 1966.

The analysis is laborious and is greatly assisted by computing machines. A simplified account of the methods used is given below. The first step in the analysis, involving the selection of events of significance to the experimenter, is known as *scanning*. The magnified image of the film is inspected and the locations of interesting events are recorded. The next stage in track analysis, known as *tracking*, involves various degrees of automation. The tracking operation is greatly expedited by utilizing a flying spot digitizer. A small spot of light is moved rapidly in this system either mechanically or electronically, across and down the image of the bubble chamber event. Each time the moving spot of light crosses a track, co-ordinates of the point are expressed in digital form. The complete co-ordinates of the event of interest are next treated by a computer to yield what is called the *geometry* of the event. For the next stage, known as *kinematic analysis*, it is required to know the masses of the particles that produce the observed tracks. The kinematic analysis, made by the computer at each vertex of an event is based on the conservation laws of momentum and of kinetic energy.

Nevertheless, cloud and bubble chambers both are important detectors of radiation from a historical point of view since they have played such an important part in the study of fundamental particles and their interaction. They are used mainly in Nuclear Physics research. It has been estimated that a comparatively small bubble chamber is equivalent in stopping power to a cloud chamber some 140 ft. long.

GEIGER-MÜLLER COUNTER

Geiger and Müller in Germany developed the counter in 1928. This counter is thus named as Geiger-Müller counter or G.M. counter. It works in Geiger region.

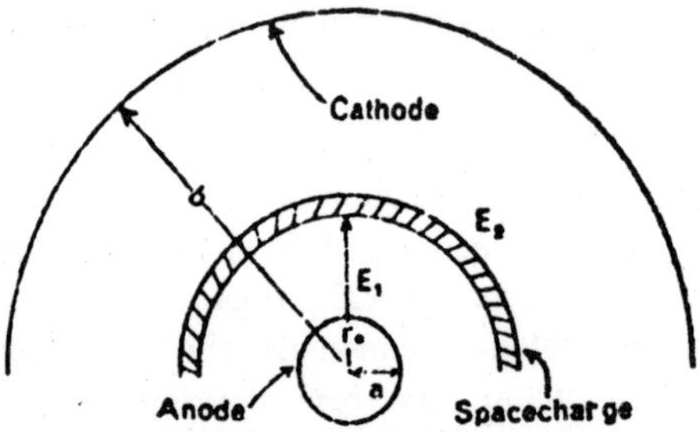

Fig. 2.21 : Positive ion sheath.

The main difference between proportional and Geiger counter is that in the *former the avalanche is formed only at one point whereas in the latter the avalanche spreads along the whole length of the central wire. The amplification does not therefore depend on the initial ionization produced by the ionizing particle.*

Pulse Formation and Decay

Let us Consider the situation when the space charge has moved outside to a radius r_0. Let there be q charges per unit length on the wire and Q charges per unit length in the space charge. The field strengths inside and outside the sheath are

$$E_1 = q/2\pi\epsilon r \text{ and } E_2 = (q + Q)/2\pi\epsilon r. \qquad ...(1)$$

Voltage applied to the counter

$$V = \int_a^{r_0} \frac{q}{2\pi \epsilon r}\, dr + \int_{r_0}^{b} \frac{q+Q}{2\pi \epsilon r}\, dr$$

Integrating we have

$$= \frac{q}{2\pi \in} \log_e \frac{b}{a} + \frac{Q}{2\pi \in} \log_e \frac{b}{r_0}$$

$$\therefore \text{ Charge } \quad q = \frac{2\pi \in}{\log_e b/a}\left[V - \frac{Q}{2\pi \in} \log_e \frac{b}{r_0}\right] \qquad ...(2)$$

The negative term in the bracket shows that q and hence E_1 is reduced during a pulse. As the positive ion sheath moves toward the cathode, r_0 increases, negative term decreases and the field strength attains its maximum initial value. The positive ion sheath will reach the cathode in about 10^{-4}sec.

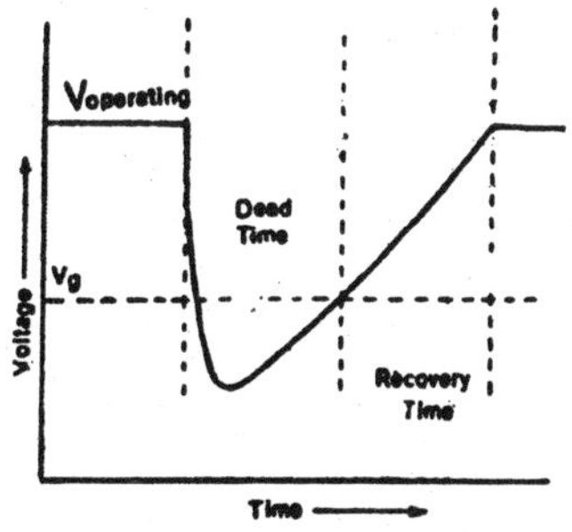

Fig. 2.22 : Dead time and recovery time.

As the positive ions reach the wall and are about 10^{-7}cm from it they cause electron to be liberated from the metal of the cathode by *field emission.* These electrons then neutralize the positive ions, leaving the excited molecules. From where they promptly go to the ground state radiating characteristic spectral lines. Some of these lines are in the ultraviolet region of the spectrum and hence liberate photoelectrons from the cathode. These secondary electrons are accelerated toward the central wire.

When the electrons reach the wire, they start a new avalanche which produces a new sheath and the discharge will continue indefinitely. This continuous discharge is not really continuous in time but consists of a series of pulses at close and some what overlapping intervals. The growth of the potential pulse at the anode is determined by the motion of the +ve ion sheath as it is repelled away from the high field region.

When the ion sheath has travelled a distance equal to about half the radius of cathode cylinder, the field close to the anode wire is below the Geiger threshold. Hence the second pulse can not be recorded. The time, during which the counter is incapable of responding to a second ionizing event, is known as *dead time* (~200μ sec.). As the space charge moves further towards the cathode, the counter sensitivity gradually returns to its original value and the second pule can now be recorded.

This time, during which pulses of reduced size are produced, is known as *recovery time.* The dead time plus recovery time may corresponds to the time which +ve ion sheath takes to reach the cathode, and is known as the *resolving time* of the counter. The high counting rate is generally reduced by reducing the high potential supply to the counter for a definite time Interval called *paralysis time.* It is achieved by an external electrical circuits generally a relaxation oscillator triggered by the step rise of Geiger pulse;

Quenching of the Discharge

The process used to prevent a continuous series of pulses from taking place in the G.M. counter are called quenching. There are three types of quenching (*suppressing*) involved in the operation of a counter:

(a) quenching of photons in the initial avalanche,

(b) electrostatic quenching of the avalanche by the +ve ion space charge, and

(c) quenching of the secondary emission when the +ve ions reach the cathode. The quenching of the discharge involves both the electrostatic quenching and the quenching of secondary emission. There are two procedures for quenching the discharge in order to improve the resolving power of G M. counters.

In the *nonself-quenched* (externally quenched) type of G. M. counter, the quenching of the discharge is achieved by means of an external resistance or by use of an auxiliary electronic circuit. In the first method counter can be quenched by using the load resistor ($10^{10}\Omega$) in series with the tube. High voltage is applied across the tube which reduces as the current pulse flows through the resistance, the gas amplification factor is thus decreased preventing a second discharge. With a chamber of 10pf capacitance this method results in a counter dead time of the order of 0.1 second. In the second method the load resistance is about $10^7\Omega$ and at the same time the voltage across the chamber is reduced, by the application of a negative voltage pulse to the anode from a monostable multivibrator, every time a count takes place. The dead time of the counting circuit can be adjusted electronically to be a chosen value.

In the *self-quenched* (internally quenched) type of G.M. counter, quenching is achieved by the inclusion of a suitable gas in the counter, such as:

(a) organic molecules, and

(b) the halogen gases. In the first case the filling is a mixture of argon and 10% of a polyatomic organic gas or vapour such as methane, ethane or ethyl alcohol as a quenching gas. As the quenching gag ionizes more readily than argon due to its lower (11.3 eV instead of 15.7 eV) ionization potential, hence the ion cloud reaching the cathode consists almost entirely of alcohol ions. These ions are neutralized by pulling electrons from the metal surface of cathode and again form largely excited molecules. *These excited molecules usually prefer decomposition rather than photon emission and hence no photons are available to start a second avalanche.* The photons if emitted by excited molecules are not likely to travel more than a millimeter through the gas before being absorbed by the photoelectric process. The process of charge exchange during collisions prevents any argon ion reaching the cathode, thereby preventing the release of any electrons from the cathode which would trigger off a second discharge.

The each transfer of ionization from argon to alcohol is accompanied by the emission of photons of energy 15.7 – 11.3 = 4.4eV. This is too low to eject an electron from a cathode.

A satisfactory quenching gas must nave three main properties:

1. When in an excited state it must prefer to dissociate rather than to de-excite by the emission of photons.
2. Its ionization potential should be lower than that of the main counting gas in the tube.
3. It must have broad and intense ultraviolet absorption bands.

The halogen quenching action seems to be similar to that of the organic agents in that some of the molecules dissociate at each discharge. However, the dissociated halogen molecules combine at the end of the discharge, so the tube has unlimited life. Halogen molecules are chemically so active that the materials of which the tube and its electrodes are made must be chosen carefully. The halogen quenched counters usually operate at 300 to 700 volts. They can be used over a temperature range of from –50°C to +60°C, whereas the organic quenched counters often cease to work at 0°C because the organic gas condenses. They are not damaged by a continuous discharge whereas poly-atomic organic gases filled counters damaged. Due to their lower voltages, long life and robustness, the halogen quenched counters are recommended for use in colleges.

In the self quenching type counters the discharge is quenched due to the internal mechanism and hence no electronic quenching circuit or high resistance is necessary. The disadvantage of high resistance is that it lengthens the recovery time of the counter. The equivalent electronic circuit permits a short time constant to be used but adds the inconvenience of an extra stage in the circuits. The advantage gained in the elimination of the quenching resistor or electronic circuit is secured at the price of stability.

G.M. Tube Construction

G. M. *tubes* have been made in a great variety of sizes and shapes, from 1 cm. to 100 cm. in length and from 0.3 cm. to 10 cm. in diameter. The walls can be of metal (copper), a metallic cylinder may be supported inside a glass tube, or the interior surface of the glass tube may be coated with a thin layer of an electrical conductor) *e.g.*, silver). The central wire is usually of tungsten with a thickness of 2×10^{-3} to 5×10^{-3}cm.

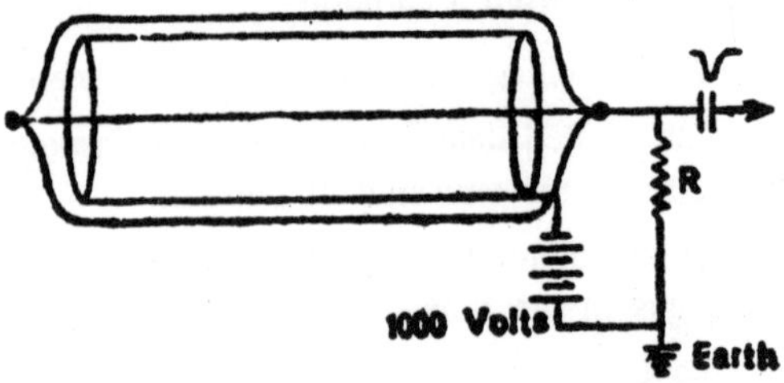

Fig. 2.23 : Basic circuit of G.M. counter.

Various observers have tried a large number of different compounds in counters and found that any polyatomic gas molecule would produce the quenching action. The molecule need not be organic. There does not appear to be any great difference in merit between various possible polyatomic molecules, each having certain advantages and disadvantages as compared to other.

For example, amyl acetate shows a some what longer useful life but a slower resolving time than does methane. To extend the life time of the tube, bromine vapour can be introduced instead of the alcohol. To detect radiation of very small penetrating power counters with very thin end windows are used. Thin end windows of mica, cellophane or glass, only 1 or 2 mg/cm^2. thick will allow a- and P-particles to pass into the counter. For detecting P-radiation thin walled glass counters (the cathode is often a thin layer of colloidal carbon deposited on the inner wall of the glass envelope), which allow the entry of β-rays from all directions, can be used. Detection of γ-rays is mainly through the ejection of photoelectron from the cathode cylinder. The probability of this process varies as Z^4. hence, counters using high Z (bismuth or lead) cathode are used for the detection of γ-rays. For detecting cosmic rays more robust counters sometimes two or three feet long are used. For our purposes:

(i) *G.M. tube should be capable of responding to the radiation to be detected,*

(ii) *It should operate at fairly low voltages,*

(iii) *It should have long life, and*

(iv) *Its cost should be minimum.*

In India G.M. tube of different shape and size are designed in Hyderabad. These are marked as I 1000, I 1002 (Helogen filled); I 1010 (Thin walled); I 1030, I 1031, I 1032 (end windows); I 1200 (liquid sample); I 1300 (gas flow detector windowless and I 1350 (gas flow detector end window).

Characteristic Curves

There are two important characteristic curves exhibited by Geiger counters. The first is the familiar *plateau curve*. This is a plot of the recorded counting rate when the voltage applied to the counter is varied. The G. M. counter, similar to proportional counter, is connected with a device capable of indicating only relatively large pulses but no small ones. Until the voltage reaches the value indicated as the *starting potential* the pulses are too small to be detected. But with rising potential, the gas amplification increases and number of pulses rise rapidly to a flat portion of the curve called *the plateau.* This is the Geiger tube region for which the *count rate is nearly independent of the potential difference across the tube.* Beyond the plateau the applied electric field is so high that a continuous discharge takes place in the tube and the count rate increases very rapidly. It should be emphasized that the shape of the knee of the plateau curve is partly a property of the associated recording circuit.

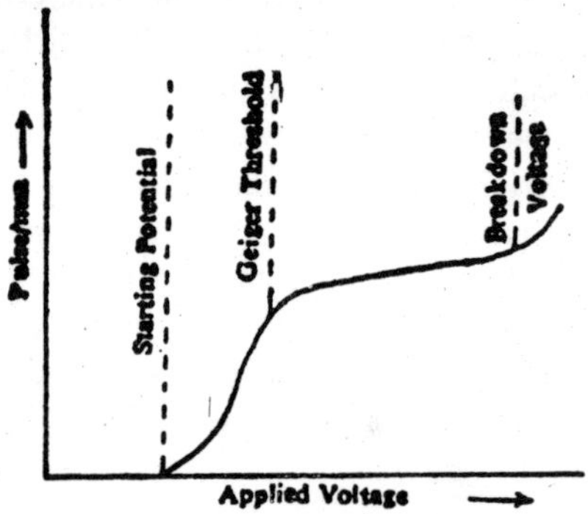

Fig. 2.24 : Plateau of G.M. tube.

$$\text{Slope of plateau } \frac{n_2 - n_1}{n_{av}} \times \frac{100}{V_2 - V_1} \times 100\% \text{ per 100 volts.}$$

The plateau has a width (or length) of a few 100 volts. A good tubes often have a plateau length of 100 to 200 volts or more with a slope of about 5% counts per 100 volts applied. *The slope for halogen filled tubes is however greater than that for organic gase filled tubes.* When the plateau becomes notably shorter and steeper, it is a sign that a counter tube is nearing the end of its useful life. The other type of curve is that of the starting potential as a function of the pressure of the gas in the counter. It can be shown from these curves that the *starting potential increases, practically linearly with the pressure of the filled gas.*

The *efficiency of the counter* is defined as the ratio of the observed counts/sec to the number of ionizing particles entering the counter per sec. Counting efficiency is defined as the ability of its counting if at least one ion pair is produced in it.

$\therefore$ Counting efficiency = $(1 - e^{-sl p})$.

where s is the specific ionization at one atmosphere, p the pressure in atmospheres and *l* the path length of the ionization particle in the counter.

Precautions in Using G.M. Tube

1. The tube must be disconnected when measurements arc over.
2. Introduction of light must be prevented in order to avoid photo electric effects.
3. The background counting rate must be subtracted.
4. The tube should never operate at a voltage higher than its normal voltage.
5. In case a continuous discharge is produced, the voltage should be decreased.
6. The high voltage must be relatively stabilized.

Main Features of a G.M. Counter

(i) Constant output pulse size, independent of initial ionization.

(ii) Sensitivity to the production of even a single ion pair,

(iii) A long insensitive time to allow entry of each particle.

(iv) Infinite life (specially with halogen quenched),

(v) Can detect α, β, γ and cosmic rays.

Pulse Counting Circuits

The pulses obtained from the G.M. tube are too small to count direct. It requires amplification, which is done frequently in two steps. The counter is immediately followed by a *preamplifier*, which has little gain but produces an output signal across a relatively low impedance. This signal is then fed into the main amplifier, which has a gain strictly independent of the size of the input pulse. The output signal is then fed into the scaling circuit.

Resolving Time and Actual Counts

Let us assume that a counting system with a resolving time τ responds at a rate n counts per unit time when exposed to N initiating events per unit time. In unit time the total insensitive time will be $n\tau$ and the number of counts missed will be $Nn\tau$.

No of counts missed = error in counting or $Nn\tau = N - n$

$\therefore$ Actual count rate $N = n/(1 - n\tau)$

Thus, the actual count rate can be calculated by knowing τ. The most usual method for it involves the measurement of four counts using two radioactive sources. The background count (counts without any source) rate B is first found. With one source, count rate is determined. Let the observed count rate be $n_1 + B$ and expected $N_1 + B$. With the first, source the second source is introduced and adjusted until the count rate is approximately doubled. We now have $N_1 + N_2 + B$ expected and $n_{12} + B$ observed. The first source is removed and the second is left in position. If the counts rate is $N_2 + B$ expected and $n_2 + B$ observed. Solving equations

$$N_1 = \frac{n_1}{(1+n_1\tau)},\ N_1 + N_2 = \frac{n_{12}}{(1-n_{12}\tau)} \text{ and } N_2 = \frac{n_2}{(1-n_2\tau)},$$

We get
$$\tau = \frac{n_1 + n_2 - n_{12}}{2n_1n_2} \quad ...(1)$$

Standard Deviation

Because of the random nature of radioactive decay it is found that if a long lived source is counted for a number of equal intervals of time the result shows considerable variations. This deviation is the square root of the average of the squares of the deviations from the mean. If N is the total number of counts in time t, then the

$$\text{Standard deviation } \sigma = \sqrt{\frac{(N)}{t}} \quad ...(2)$$

$$\text{The coefficient of Variation } V = \frac{\text{Standard deviation } \sigma}{\text{Rate of counting R}}$$

$$= \frac{\sqrt{(N)/t}}{N/t} = \frac{1}{\sqrt{(N)}} \quad ...(3)$$

Account must be taken of the count rate due to background, unless the count rate of the sample is very large. Hence actual counting rate

$$RT = R_s + R_B,$$

where the subscripts refer to the total, sample and background respectively.

The standard deviation $\sigma s = \sqrt{(\sigma T^2 + \sigma n^2)}$

or $$\sigma s = \sqrt{\left(\frac{N_T}{tT^2} + \frac{N_B}{t_B{}^2}\right)} = \sqrt{\left(\frac{R_T}{tT} + \frac{R_B}{t_B}\right)} \quad ...(4)$$

and $$V_S = \frac{\sigma s}{Rs} = \frac{\sigma s}{R_T - R_B} \quad ...(5)$$

SCINTILLATION COUNTERS

One of the earliest methods of detecting nuclear radiations was by the luminescence, they produced in certain substances. In 1903, Crookes in England and Elster and Geitel in Germany independently reported that alpha particles impinging on zinc sulphide screens produced individual flashes which could be observed through a microscope. The first attempt to count α-particles, by observing flashes of light (scintillations) generated in diamond by them, was made by E. Regener in Germany in 1908. In the same year *Rutherford and Geiger compared the number of scintillations seen with the number of pulses recorded by an ionization chamber. The numbers were same in both cases. Hence, if each particle caused a single pulse in the counter then it also gave rise to one scintillation.* In this way this method was available for counting individual α-particle. The observations of scintillations through low power microscope were very tedious and limited to relatively low counting rates. For this reason early forms of scintillation method soon became obsolete and more attention was paid to the development of gas counters.

Since 1945, this old technique, is now playing a most important role. This has been due to the development of the photo-multiplier tube and the intensive study of the luminescent properties of many inorganic solids as well as organic compounds either in the solid state or in solution. A purely schematic diagram of a scintillation counter unit is shown in the following figure. A scintillation counter consists of the luminescent material (known is *scintillator*), reflecting layer such as aluminium foil enclosing the luminescent substance to facilitate the collection of light, light pipes, the photomultiplier tube, amplifier, voltage discriminator and an electrical circuit to record the output pulses. Before considering the performance of the scintillation counter let us first examine the behaviour of some of its components.

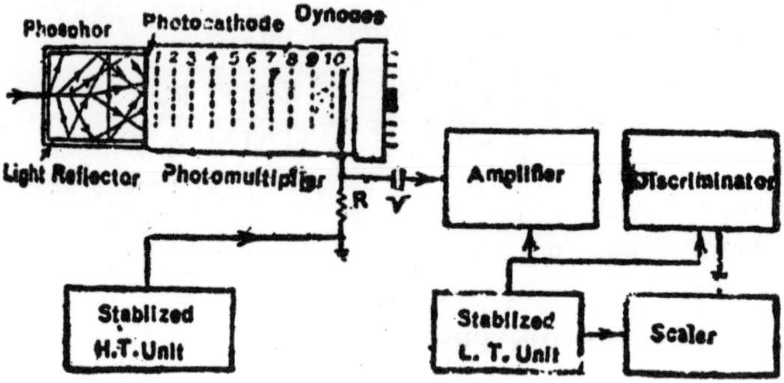

Fig. 2.25 : Schematic diagram of Scintillation Counter.

Scintillators

The principal methods of excitation are *incandescence and luminescence.* The luminescence caused by photons are named as photo luminescence and are further divided into phosphorescence and fluorescence. When the visible or ultraviolet light is emitted within 10^{-8}sec or less after the radiation absorption the emission is called fluorescence.

Phosphorescence refers to delayed light emission, which may follow the radiation absorption by minutes, day or even years. For the moment we shall consider fluorescence response. The substances scintillate when bombarded by radiations are named as scintillators and are mainly classified as:

(i) Organic Liquids

Some solutions of organic compounds are also capable of acting as phosphors, although, they are not as efficient as the best pure crystals. The liquid scintillator has two main components: the solvent, (*e.g.*, toluene or xylene) and the primary solute (*e.g.*, a few percent of diphenyloxazole or terphenyl). Most of the nuclear radiation energy is absorbed by the solvent which is not a scintillator and is then transferred to the solute which emits the light actually. When the scintillations occur in the extreme ultraviolet region one uses a light shifter to increase the wavelength of the emitted light.

(ii) Organic Crystal

Kallmann employed a large clear crystal of the organic substance naphthalene as phosphor. Following this, it has been found that other related organic compounds, consisting of several linked benzene rings are better scintillators. Of the solid organic scintillators anthracene and stilbene appear to be the best. The organic crystalline scintillators have a faster decay time ($\sim 10^{-8}$ sec), high transparency and poorer conversion efficiency especially for heavy particles. For β-particles response is linear. Because of low Z and ρ, these crystals have low cross-sections for photoelectric and pair production processes, hence can not be used for the detection γ-rays and X-rays.

(iii) Inorganic Crystals

Unlike the organic compounds, inorganic substances do not scintillate when they are pure. Best one, especially for gamma rays, is crystalline *sodium iodide activated with thallium* (about 0.1 per cent). A less sensitive alternative is the nonhygroscopic cesium iodide, also activated with thallium. Calcium iodide with a trace of europium is more efficient than NaI, although, it is also hygroscopic. Silver activated zinc sulphide is an excellent scintillator but can be used only in thin layers, because it soon becomes opaque to the light emitted by it.

(iv) Plastics

Organic scintillator system, which lies between a solid and liquid has a plastic as the base. A solution is made of the primary solute in a solvent like vinyl toluene or styrene which can be readily converted into a solid plastic. The resulting transparent material can be made in large pieces and cut to any desired shape. Plastic scintillators are widely used in high energy physics.

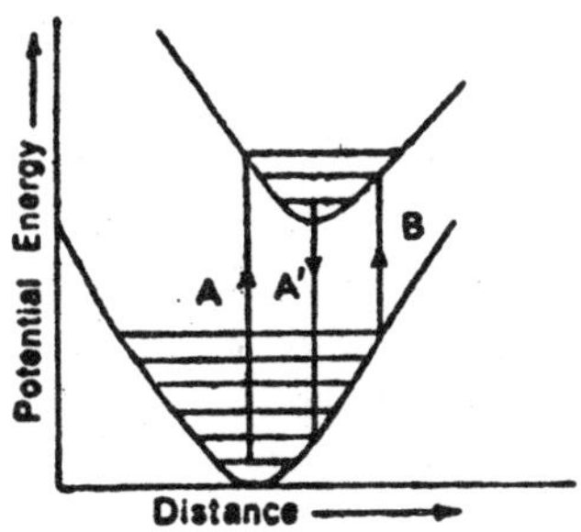

Fig. 2.26 : Luminescence proccess in NaI (Tl).

In a crystal such as NaI (T*l*) an ion Tl^+ replaces an ion Na^+ and forms an emission center. The T*l* ion thus occupies a position of equilibrium in the lattice such that a displacement from that position produces an increase in potential energy for the entire lattice system near the Tl^+ ion.

The absorption of radiation changes the Tl^+ ion to an excited state for which the potential energy curve has a minimum at a position different from that in the ground state. The absorption (transition A) leaves the system in a state of high vibrational energy.

The excited particle makes many collisions and with in 10^{-5} to 10^{-8} sec dissipates its surplus vibrational energy to the lattice. The most intense emission thus corresponds to transition A'. If the temperature is high enough, so that the higher vibrational levels of the ground state have appreciable population. Now, it is possible that the absorption takes place along B which is followed by emission of radiation.

ZnS (activated with Cu or Ag)

1. The output is proportional to the energy of the incident particle.
2. C_{IP} is independent of the incident energy.
3. It has highest light conversion efficiency C_{IP} of any known phosphor.
4. The efficiency and emission spectra depend on the preparation and activator used.

For ZnS (activated with Cu)—conversion efficiency $C_{IP} = 28\%$ and the emission spectra from 4000-6000Å has a peak at 4500Å.

For ZnS (activated with

Cu)— $C_{IP} = 25\%$ and $\lambda_{peak} = 5200$Å.

It has following limitations

1. Decay time is ~10^{-6} sec or longer.
2. A large crystal can not be prepared.
3. It has low transparency. Beyond thickness 25 mg/cm^2, the detector efficiency decreases.
4. It can not be used for the detection or γ-rays and electrons. Only the detection of slightly heavier particles such as a, p, d is possible.

NaI (activated with Tl)

1. Its absorption spectra lies in ultraviolet region, with the peaks at 2340 and 2930Å. Due to this reason, the crystals are highly transparent.
2. Its emission spectra has a average value of wavelength ~4100Å, and FWHM = 850 ± 100Å.
3. Its can be grown into large single crystal.
4. It has apart from ZnS, the highest.
5. Because of high values of p and Z, it has large cross sections for pair-production, photoelectric effect and compton effect. Due to this reason it is very suitable for the detection of gamma and X-rays.
6. The decay time at room temperature is lowest ($\sim 2.5 \times 10^{-7}$sec.).

C_{IP} decreases with increasing specific ionization causing a nonlinear response to α-particles. The response is linear to p, d and e. The crystal must be protected from moisture with surface film of liquid paraffin or be placed in dry atmosphere.

(e) Gases

Inert gases of the atmosphere, *e.g.*, argon, krypton and xenon when exposed to nuclear radiations produce scintillations in the ultraviolet region of the spectrum. To increase photomultiplier response the flashes are converted into visible light by means of a suitable wavelength shifter. An important advantage of the inert element scintillators iff that the light output is proportional very closely to the energy of the nuclear radiations.

Choice of Phosphors

NaI (Tl) is a highly effective scintillator as far as light output is concerned, but the light pulse has a relatively high decay time, compared with the organic phosphors. Hence *NaI (Tl) is less satisfactory when a very rapid response is required.* Iodine present in it gives a much greater stopping power for γ-rays. The size of the output pulse is proportional to the energy.

Organic scintillators give comparatively smaller light pulses, but have a shorter decay time. Hence, a higher counting rates are possible. The decay time in the liquid and plastic scintillators are even shorter than in the crystals Because of the small stopping power, the organic scintillators cannot be used for γ-rays unless the counter is large and are used for β-detection.

Screens with activated ZnS, in very thin layers, are useful for the detection of α-particles. Neutrons can be detected through the emission of a suitable charged particle resulting from a nuclear reaction initiated by the incident neutron.

Different Processes in the Scintillation Detector

(a) Absorption Process

If the dimensions of the scintillator are large compared to the range of the incident particle, the incident charged particle can dissipate all its energy to the scintillator. If the γ-ray is incident on a scintillator it may interact with the scintillator material in three ways: *photo-electric absorption, Comption scattering and pair production* each cases, electrons are produced. The all or part of the energy of incident γ-ray is transformed into the kinetic energy of these electrons.

(b) Scintillation Process

The electrons emitted will give up their kinetic energy in ionization or excitation in the scintillator material. This absorbed energy appears either as heat energy or as luminescence photons. In the latter cases the scintillator material de-excites by light emission within 10^{-8} seconds. This emitted light is known as scintillation. If τ is the mean life of the scintillator, the number of light photons in time t after the ionization radiation has arrived is given by

$$n = \text{const } (1 - e^{-t/\pi}) \qquad ...(1)$$

(c) Formation of Electrical Pulses

The light produced in scintillator falls on the photocathode of the photomultiplier tube, producing photoelectrons. These electrons fall on the first dynode, producing a bunch of secondary electrons. This process is repeated and a gain upto $10^{-7} - 10^{8}$ is achieved. The electrons finally fall on the electrode, known as anode, and thus, produce an electrical pulse across the load resistance R. The pulse height V at the output related to the amount of charge Q collected at the output through the relation

$$V = \frac{Q}{C} \quad ...(2)$$

where C is the capacity of the output point.

If E_i be the energy of the incident charged particle or γ-rays and A the probability of amount of energy absorbed by the phosphor.

∴ Energy dissipated in the phosphor = E_iA.

This energy is converted within efficiency C_{IP} into photons of average energy E_p.

$$nr = E_iA\ C_{IP}/E_p \quad ...(3)$$

If T_p is the optical transparency and G is the fraction of the photons falling on the photocathode. Hence, the number of photons incident on the photocathode.

$$n_p/ = T_pGn_p = E_iAC_{IP}T_pG/E_p.$$

These photons are converted into photoelectrons with an efficiency C_{PE} f(v), where C_{PE} is the photoelectric conversion efficiency of the cathode and f(v) is the relative response at frequency v.

∴ No. of photo electrons $n = C_{PE}/(v)\ np/$

If f_d be the fraction of photoelectrons reaching the first dynode and M the electron multiplication factor, then the total number of electrons reaching the anode

$$N = f_dMn = C_{PE}f(v)\ E_iAC_{IP}T_pGf_dM/E_p.$$

Hence total charge Q = eN and the pulse height

$$V = Q/C = eE_i\ AC_{IP}T_pGC_{PE}f(v)f_dM/CE_p \quad ...(4)$$

Here, we see that the pulse height is proportional to the incident particle if the particle is absorbed in the scintillator completely.

Resolving Power

The height of the output pulse exhibits variations about the mean value for a given particle energy E. This variation is due to the statistical fluctuations in the various factors of eqn. (4). The resolution R of a scintillation counter can be defined as follows

$$\frac{1}{R} = \frac{\overline{Q}^2 - (\overline{Q}^2)}{(\overline{Q}^2)} \quad ...(5)$$

where $\overline{Q}$ is the mean charge collected at the anode and $\overline{Q}^2$ is the mean squared charge collected at the anode. The smaller the half width of the photo peak, the higher will be energy resolution R.

The scintillation counter can be used as a spectrometer in conjunction with an electronic pulse height analyzer, which has found many applications in low energy, as well as in high-energy nuclear physics. Suitable circuits have been designed to sort out particles with in a given range of energy so that distribution in energy of the incident radiation may be measured.

Unfortunately, even mono-energetic γ-rays produce a complex pulse height spectrum. The pulse height spectrum produced by γ-rays from the lower excited states of Mg^{24} is shown in Fig. 1.24 (right). This isotope is produced in low lying states by β-emission of Na^{24} according to the decay scheme shown in Fig. 2.27 (left).

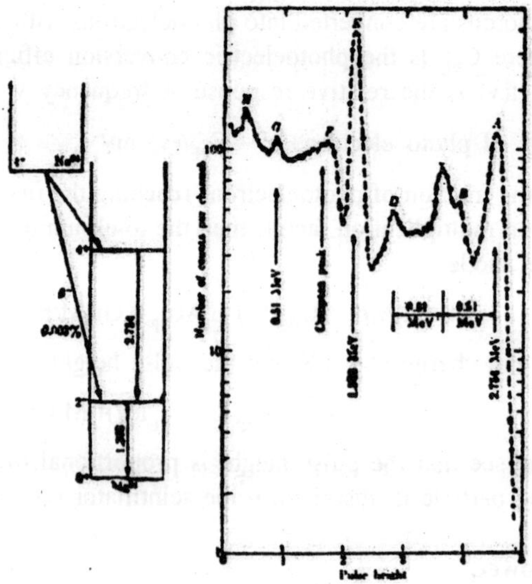

Fig. 2.27 : Decay scheme for Na^{24} (Left). Pulse height spectrum for gamma rays (Right).

Two γ-rays of energy, one at 2.754 MeV and other at 1.368 MeV are emitted. The peak A in this spectrum corresponds to the full energy of 2.754 γ-ray and is due to pair production process only.

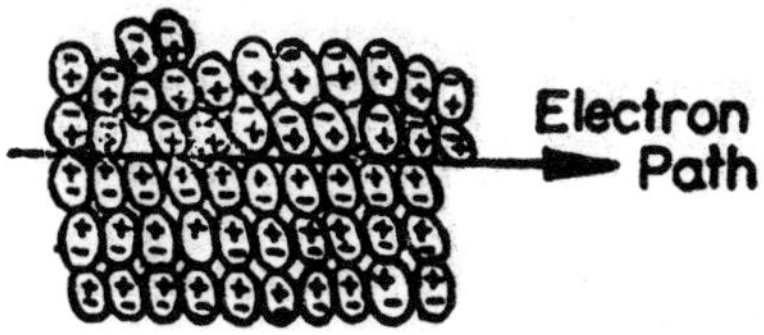

Fig. 2.28 : Polarization of molecules.

The next peak B corresponds to the electrons produced by compton process. The peak C is caused by pair-production processes in which one of the two annihiliation photons has escaped and peak D corresponds to the processes in which both the annihiliation photons have escaped. The central sharp peak E corresponds to the energy 1.368 MeV and is due to pair production or photo electric process. The compton peak corresponding to this peak is F. As 1.368 MeV pair production is relatively unimportant, hence first and second escape peaks corresponding to the energies 1.368 – 0.51 = .858 and 1.368 – 1.02 = .348 are not observed. Peak G corresponds to 0.51 MeV annihiliation photons produced outside the scintillator but absorbed by it. The peak H arises from back scattered photons from compton events.

The scintillation counter has a much higher counting efficiency for gamma rays due to the greater amount of energy dissipation by the gamma rays. The scintillation counter offers greater stability, greater accuracy, shorter resolving time and higher efficiency than G.M. counter. Provided all the particle energy is dissipated in the scintillator, the voltage pulse height produced can be used to measure the particle energy, once the counter is calibrated by means of a standard source. Because of their extremely rapid response scintillation counters are having many applications in atomic energy field. These are used for the accurate timing of nuclear particles moving with very high speeds.

SOLVED EXAMPLES

Example 1:

An organic-quenched G.M. tube operates at 1000 volts and has a wire diameter of 0.2 mm. The radius of the cathode is 2 cm, and the tube has a guaranteed life-time of 10^9 counts. What is the maximum radial field and how long will the counter last if it is used on the average, for 60 hours per week at 2000 counts per minute?

Solution:

The maximum radial field or the field along a radius at the wire surface

$$E_{maz} = \frac{V_0}{2.3026x \log_{10} (b/a)} = \frac{1000}{2.3026 \times 10^{-4} \times \log_{10} 200}$$

$$= 1.8^9 \times 10^6 \text{ Volts/metre.}$$

If the lifetime of the tube is N years the total number of counts recorded will be equal to the guaranteed counts. Hence

$$N \times 50 \times 60 \times 60 \times 2000 = 10^9$$

or $N = 2.77$ years.

Example 2:

Estimate the number of ion pairs produced in a proportional counter by a 10 MeV proton if the counter size and pressure are large enough for all the proton's energy to hc absorbed. If the gas multiplication factor is 10^3, how many coulombs flow in the counter when this proton is absorbed? If the pulse of current flows for about 10^{-3} sec and if the resistance is 10^4 ohms, estimate the height of voltage pulse.

Solution:

Energy required for the production of one ion pair = 35 eV, hence the no. of ion pairs produced by a proton of energy 10 MeV

$$= 10 \times 10^6/35 = 2.86 \times 10^5$$

After amplification the total number of ion pairs = nM

$$= 2.86 \times 10^5 \times 10^3 = 2.86 \times 10^8.$$

Charge carried by these ions $\Delta Q = 2.86 \times 10^8 \times 1.6 \times 10^{-19}$

$$= 4.57 \times 10^{-11} \text{ coulomb.}$$

Hence the current passes through the resistor ΔI

$$= 4.57 \times 10^{-11}/10^{-3} = 4.57 \times 10^{-8} \text{ amp.}$$

$\therefore$ The height of the pulse $\Delta V = I.R. = 4.57 \times 10^{-8} \times 10^4$

$$= 4.57 \times 10^{-4} \text{ volt.}$$

Example 3:

Compute the approximate number of quanta of visible light emitted as Cerenkov radiation in the frequency range corresponding to wavelengths in vacuum of λ = 4000 to 7000 A.U. given off by an electron of energy 20 MeV traversing 1 cm of Incite (μ 1.49).

Solution:

Relativistic kinetic energy is given by $E = m_0c^2\ [(1-\beta^2)^{1/2} - 1]$

$$\frac{1}{\sqrt{(1-\beta^2)}} - 1 = \frac{E}{m_0c^2} = -\frac{20 \times 1.6 \times 10^{-12}}{9.109 \times 10^{31} \times (3 \times 10^8)^2}$$

$\therefore\quad (1-\beta^2)^{-1/2} = 40$ or $\beta^2 = 0.9995$

Hence the number of quanta emitted as Cerenkov radiation per metre

$$N = \frac{2\pi z^2}{137}\left(\frac{1}{\lambda_2} - \frac{1}{\lambda_1}\right)\left(1 - \frac{1}{\beta^2\mu^2}\right)$$

$$= \frac{2 \times 3.14}{137} \times \frac{3000 \times 10^{+10}}{4000 \times 7000} \times \left[1 - \frac{1}{0.9995 \times (1.49)^2}\right]$$

$= 26980$ quanta/metre

$= 270$ quanta/cm (Approx.)

Example 4:

A.G.M. tube with a cathode 5.0 cm. in diameter and a wire diameter of 0.012 cm is filled with argon to a pressure such that the mean free path is 7.8 × 10^{-4} cm. Calculate the value of the voltage that must be applied to just produce an Avalanche.

Solution:

Electric field strength E at any radius r between two concentric cylinders, when a potential V is applied is given by

$$E = \frac{V}{r \log_e (b/a)}$$

Avalanche will just, start when the integral of this equation, taken over one mean free path is equal to the ionisation potential of the filling gas, *i.e.*,

$$E_{a\lambda} = \frac{V\lambda}{a \log_e (b/a)} = I \text{ or } \frac{Ia \log_e (b/a)}{\lambda}$$

$$V = \frac{15.7 \times 0.006 \times 10^{-2} \times 2.3026 \log_{10} (0.025/0.00006)}{7.8 \times 10^{-6}}$$

$$= 730 \text{ volts.}$$

Example 5:

A Geiger-Müller counter has a plateau slope of 3% per 100 volts. If the operative point is at 1100 volts, what is the maximum permissible voltage fluctuation if the counting is not to be affected by more than 0.1?

Solution:

As the plateau slope $= \frac{n_2 - n_1}{n_{av}} \times \frac{100}{V_1 - V_1} \times 100\%$ per 100 volts

In this problem $(n_2 - n_1)/n_{av} = 0.1\% = 0.1 \times 10^{-2}$ and slope = 3

$\therefore V_2 - V_1 = 0.1 \times 10^{-2} \times 1100 \times 100/3 = 3.3$ volts.

As operating voltage is 1100 volts, hence the voltage fluctuation is 3.3 volts in 1100 volts or 0.3%.

Example 6:

A slow neutron entering a U^{235} fission chamber releases fission fragments of a total energy of 200 MeV. If the capacity of the collector system is 25 μμf, calculate the resultant pulse height.

Solution:

Since the energy required for the production of one ion pair is 35 eV, hence the total number of ion pairs produced

$$N = 200 \times 10^6/35 = 57.14 \times 10^5$$

$\therefore$ Total charge produced = Ne = $57.14 \times 10^5 \times 1.6 \times 10^{-19}$ coul.

$= 91.42 \times 10^{-14}$ coul.

Pulse height $\Delta V = \frac{\Delta q}{C} = \frac{91.42 \times 10^{-14}}{25 \times 10^{-12}} = 3.65 \times 10^{-2}$ volt.

Example 7:

An anthracene crystal and a 10 stage photomultiplier tube are to be used as a scintillation spectrometer. It is desired that a 10 keV β-particle incident on the scintillator produce a 2 mV pulse on the photo-multiplier output circuit, which has a capacity of 120 μμf. What average electron multiplication per stage is required in the photomultiplier? Assuming a light collection efficiency unity and a photocathode efficiency 0.1.

Solution:

Anthracene gives the highest yield of photons, about 15 for each 1000 eV of energy dissipated in the crystal. Hence, the number of photons emitted in our problem is 150. Since the light collection efficiency is unity and the photocathode efficiency is 0.1, hence number of photoelectrons emitted is 15. If each dynode produces n secondary electrons, after passing through 10 stages the electrons are multiplied by a factor n^{10}.

∴ No. of electrons in the output = $15 \times n^{10}$,

Hence the output pulse $\Delta V = \frac{\Delta q}{C} = \frac{15 \times n^{10} \times 1.6 \times 10^{-19}}{120 \times 10^{-12}}$

As it is given to be 2mV, hence

$$n^{10} = \frac{120 \times 10^{-12} \times 2 \times 10^{-3}}{15 \times 1.6 \times 10^{-19}} = 10^5$$

or $$n = 3.16 \text{ (Approx.)}.$$

EXERCISES

1. Scintillator with a background count rate of 260 cpm is used to assay samples with average 900 cpm. What will be the coefficient of variation and the standard deviation of an, "equal-count" assay of 10,000 counts each ? What time will be required for each complete determination.
2. A sample of uranium, emitting a-particles of energy 4.18 MeV, is placed near an ionization chamber. Assuming that only 10 particles per sec. enter the chamber, calculate the current produced.

3. A. 0.1 MeV electrons is absorbed in an anthracene scintillator accompanied by the 10 stage photomultiplier. If it produces a 0.01 V pulse in the photomultiplier output circuit, which has a capacity of 160 μμf. What average electron multiplication per stage is required in the photomultiplier. Assume a photocathode efficiency of 0.07.
4. A G.M. tube with a cathode diameter of 4.0 cm, and a wire diameter of 0.016 cm is filled with argon and alcohol to a pressure such that the mean free path is 4.6×10^{-3} cm. Calculate the maximum radius at which secondary ions will be formed when 1200 volts is applied to the tube.
5. A silicon detector has an area of 3.0 cm^2 and a depletion layer of 60 microns when polarised with 50 volts. Calculate the time constant of the circuit when used with a 100 megaohm resistance, the time required to collect all of the holes and the amplitude of the voltage pulse produced by the absorption of a 3.78 MeV α-particles.
6. Calculate the resolving time of a pulse counter from the following schcdule of count rates: background 65 cpm, first source in position 8350 cpm, both sources in position 16832 cpm, second source in position 9416 cpm.
7. The paralysis time of a G.M. tube is 400m sec. What is the true count rate for measured count rates of 100, 1,000 and 10,000 counts per min.

3

Nuclear Transmutation and Reactions

INTRODUCTION

In this section reference will be made to specific cases of nuclear reactions with various projectiles.

1. Transmutation by Alpha Particles

(a) **(α-p) Reaction:** Rutherford first observed in 1919 the transmutation of the nitrogen nucleus by α-particles in which α-particle was captured by a nucleus forming a compound nucleus which disintegrated immediately into a new nucleus by the ejection of a proton. Few of these (α-p) reactions are :

(i) ${}_{21}Sc^{45} + {}_{2}He^{4} \rightarrow ({}_{23}V^{49})^{*} \rightarrow {}_{22}Ti^{48} + {}_{1}H^{1} - 0.3$ MeV.

(ii) ${}_{19}K^{39} + {}_{2}He^{4} \rightarrow ({}_{21}Sc^{43})^{*} \rightarrow {}_{20}Ca^{42} + {}_{1}H^{1} - 0.89$ MeV

(iii) ${}_{16}S^{32} + {}_{2}He^{4} \rightarrow ({}_{18}A^{36})^{*} \rightarrow {}_{17}Cl^{35} + {}_{1}H^{1} - 2.10$ MeV

(iv) ${}_{14}Si^{28} + {}_{2}He^{4} \rightarrow ({}_{16}S^{32})^{*} \rightarrow {}_{15}P^{31} + {}_{1}H^{1} - 1.92$ MeV

(v) ${}_{13}Al^{27} + {}_{2}He^{4} \rightarrow ({}_{15}P^{31})^{*} \rightarrow {}_{11}Si^{30} + {}_{1}H^{1} + 2.26$ MeV

(vi) ${}_{9}F^{19} + {}_{2}He^{4} \rightarrow ({}_{11}Na^{23})^{*} \rightarrow {}_{10}Ne^{22} + {}_{1}H^{1} + 1.58$ MeV

(vii) ${}_{5}B^{10} + {}_{2}He^{4} \rightarrow ({}_{7}N^{14})^{*} \rightarrow {}_{6}C^{13} + {}_{1}H^{1} + 4.04$ MeV

Few (α-p) reactions lead to the formation of radiative isotopes that decay by β-emission. For example

$$_5B^{11} + _2He^4 \rightarrow (_7N^{15})^* \rightarrow _6C^{14} + _1H^1 + 0.75 \text{ MeV}$$

followed by $_6C^{14} \rightarrow _7N^{14} + \beta, - T = 5580$ years.

(b) (α-n) Reaction : The bombardment of beryllium by α-particle with the subsequent emission of neutrons is one of the (α-n) reactions. Few of nuclear reactions known as (α-n) type are :

(i) $_{13}Al^{27} + _2He^4 \rightarrow (_{15}P^{31})^* \rightarrow _{15}P^{30} + _0n^1$.

(ii) $_9F^{19} + _2He^4 \rightarrow (_{11}Na^{23})^* \rightarrow _{11}Na^{22} + _0n^1$

(iii) $_7N^{14} + _2He^4 \rightarrow (_9F^{18})^* \rightarrow _9F^{17} + _0n^1$

(iv) $_5B^{11} + _2He^4 \rightarrow (_7N^{15})^* \rightarrow _7N^{14} + _0n^1$

(v) $_4Be^9 + _2He^4 \rightarrow (_6C^{13})^* \rightarrow _6C^{12} + _0n^1$

(vi) $_3Li^7 + _2He^4 \rightarrow (_5B^{11})^* \rightarrow _5B^{10} + _0n^1$.

In many of these reactions the product nuclei are left in excited states. Most of the product nuclei formed by (α-n) reaction are unstable isotopes which then disintegrate with the emission of a positron. Few of these are given below with their half lives.

(i) $_{15}P^3 \rightarrow _{14}Si^{30} + \beta^+$, T = 2.5 min.

(ii) $_{14}Si^{28} \rightarrow _{13}Al^{27} + \beta^+$, T = 4.9 sec.

(iii) $_{13}Al^{26} \rightarrow Mg^{26} + \beta^+$, T = 7 sec.

(iv) $_{11}Na^{22} \rightarrow _{10}Ne^{22} + b^+$, T = 2.6 years.

(v) $_9F^{17} \rightarrow _8O^{17} + \beta^+$, T = 1.2 min.

(vi) $_7N^{13} \rightarrow _6C^{13} + \beta^+$, T = 10.1 min.

(c) When the energy of the a particle is large enough, two or more nucleons may be expelled in a nuclear reaction, processes of the (α, 2n), (α, np), (α, 3n), (α, 4n), (α, 3np), etc., types have been detected.

(d) Radiative Capture : It is also possible, for the compound nucleus formed by the capture of an α-particle to go to a more stable configuration without emitting a particle but with emitting a γ-ray photon.

$$_3Li^7 + _2He^4 \rightarrow (_5B^{11}) \rightarrow _5B^{11} + h\nu.$$

2. Transmutation by Protons

(a) (p-α) Reaction : The first artificial transmutation produced by protons was given by Cockcroft and Walton (1932). The nuclear reaction is given by equation

$_3Li^7 + {}_1H^1 \rightarrow ({}_4Be^8)^* \rightarrow {}_2He^4 + {}_2He^4 + 17.28$ MeV.

Other examples are :

(i) $_{17}Cl^{35} + {}_1H^1 \rightarrow ({}_{18}A^{36})^* \rightarrow {}_{16}S^{32} + {}_2He^4$

(ii) $_{11}Na^{23} + {}_1H^1 \rightarrow ({}_{12}Mg^{24})^* \rightarrow {}_{10}Ne^{20} + {}_2He^4$

(iii) $_9F^{19} + {}_1H^1 \rightarrow ({}_{10}Ne^{20})^* \rightarrow {}_8O^{16} + {}_2He^4$

(iv) $_5B^{11} + {}_1H^1 \rightarrow ({}_6C^{12}) \rightarrow {}_4Be^8 + {}_2He^4$

(v) $_4Be^9 + {}_1H^1 \rightarrow ({}_5B^{10})^* \rightarrow {}_3Li^6 + {}_2He^4$

(vi) $_3Li^6 + {}_1H^1 \rightarrow ({}_4Be^7)^* \rightarrow {}_2He^3 + {}_2He^4$

(b) **(p-n) Reaction :** The probability of the (p, n) reaction is relatively high if the proton has sufficient energy to penetrate the Coulomb energy barrier. Few examples are :

(i) $_{29}Cu^{35} + {}_1H^1 + \rightarrow ({}_{30}Zn^{66})^* \rightarrow {}_{30}Zn^{65} + {}_0n^1$

(ii) $_{28}Ni^{58} + {}_1H^1 \rightarrow ({}_{29}Cu^{59})^* \rightarrow {}_{29}Cu^{58} + {}_0n^1$

(iii) $_{11}Na^{23} + {}_1H^1 \rightarrow ({}_{12}Mg^{24})^* \rightarrow {}_{12}Mg^{23} + {}_0n^1$

(iv) $_8O^{18} + {}_1H^1 \rightarrow ({}_9F^{19})^* \rightarrow {}_9F^{18} + {}_0n^1$

(v) $_5B^{11} + {}_1H^1 \rightarrow ({}_6C^{12})^* \rightarrow {}_6C^{11} + {}_0n^1$.

In these reactions, the product nucleus is nearly same as the target nucleus but its atomic number is one unit higher. These equations are usually endoergic.

(c) If the incident proton has an energy (> 20 MeV) the compound nucleus has sufficient excitation energy to permit the expulsion to two or more nucleons. Few examples are:

Cu^{63}(p, pn) C^{62}_{11}, Cu^{63} (p, 2d) Zn^{62} and Sr^{88} (p, 3n) Y^{86}.

(d) **(p-d) Reaction :** It is an another type of disintegration caused by protons Examples of this type of reaction are :

$_3Li^7 + {}_1H^1 \rightarrow {}_3Li^6 + {}_1H^2$

$_4Be^9 + {}_1H^1 \rightarrow {}_4Be^8 + {}_1H^2$.

Reactions of this type are more common with heavier elements and protons of fairly high energy. They appear to be pick-up reaction and do not form any compound nucleus.

(e) **Proton Capture :** In some cases the compound nucleus is in excited state which returns to its normal state with the gamma ray photon. Examples of this reaction are :

$$_3Li^7 + {}_1H^1 \rightarrow ({}_4Be^8)^* \rightarrow {}_4Be^8 + \gamma$$
$$_6C^{12} + {}_1H^1 \rightarrow ({}_7N^{13})^* \rightarrow {}_7N^{13} + \gamma$$
$$_7N^{14} + {}_1H^1 \rightarrow ({}_8O^{15}) \rightarrow {}_8O^{15} + \gamma$$
$$_9F^{19} + {}_1H^1 \rightarrow ({}_{10}Ne^{20})^* \rightarrow {}_{10}Ne^{20} + \gamma$$
$$_{13}Al^{27} + {}_1H^1 \rightarrow ({}_{14}Si^{28})^* \rightarrow {}_{14}Si^{28} + \gamma$$
$$_{24}Cr^{50} + {}_1H^1 \rightarrow ({}_{25}Mn^{51})^* \rightarrow {}_{25}Mn^{51} + \gamma$$

3. Transmutation by Neutrons

Neutrons have no electric charge and can penetrate positively charged nuclei without any experience of repulsive electrostatic forces, hence neutrons have also been found to be extremely effective.

(a) **(n-p) Reaction :** In this reaction the proton in the nucleus is replaced by a neutron. The mass number is not changed but the charge is decreased by one unit. Few examples of (n-p) reaction produced by slow neutrons are :

$$_7N^{14} + {}_0n^1 \rightarrow ({}_7N^{15})^* \rightarrow {}_6C^{14} + {}_1H^1 + Q$$
followed by $_6C^{14} \rightarrow {}_7N^{14} + \beta^- + E_{max}$.
$$_2He^3 + {}_0n^1 \rightarrow ({}_2He^4)^* \rightarrow {}_1H^3 + {}_1H^1 + Q$$
followed by $_1H^3 \rightarrow {}_2He^3 + \beta^- + E_{max}$.
$$_{17}Cl^{35} + {}_0n^1 \rightarrow ({}_{17}Cl^{36})^* \rightarrow {}_{16}S^{35} + {}_1H^1 + Q$$
followed by $_{16}S^{35} \rightarrow {}_{17}Cl^{35} + \beta^- + E_{max}$

Other (n-p) reactions performed with fast neutrons have Q-values negative. Few examples are :

$$_{13}Al^{27} + {}_0n^1 \rightarrow ({}_{13}Al^{28})^* \rightarrow {}_{12}Mg^{27} + {}_1H^1$$
$$_{16}S^{32} + {}_0n^1 \rightarrow ({}_{16}S^{33})^* \rightarrow {}_{15}P^{32} + {}_1H^1$$
$$_{29}Cu^{65} + {}_0n^1 \rightarrow ({}_{29}Cu^{66})^* \rightarrow {}_{28}Ni^{65} + {}_1H^1$$
$$_{30}Zn^{64} + {}_0n^1 \rightarrow ({}_{30}Zn^{65})^* \rightarrow {}_{29}Cu^{64} + {}_1H^1.$$

In all these reactions the product nuclei are radioactive producing finally nuclei identical with the initial target nuclei.

(b) **(n-α) Reaction :** In this type of reactions, the capture of slow neutron results in the emission of an α-particle. Few examples are :

$$_5B^{11} + {}_2He^4$$

(i) $_7N^{14} + {_0n^1} \rightarrow (_7N^{15})$

$_3Li^7 + {_2He^4} + {_2He^4}$.

(ii) $_{13}Al^{27} + {_0n^1} \rightarrow (_{13}Al^{28})^* \rightarrow {_{11}Na^{24}} + {_2He^4}$.

(iii) $_5B^{10} + {_0n^1} \rightarrow (_5B^{11})^* \rightarrow {_3Li^7} + {_2He^4}$

(iv) $_3Li^6 + {_0n^1} \rightarrow (_3Li^7)^* \rightarrow {_1H^3} + {_2He^4}$

The capture of fast neutrons by heavier nuclei results in the emission of an alpha particle, the product nucleus is usually radioactive. Some examples are :

$_{11}Na^{23} + {_0n^1} \rightarrow (_{11}Na^{24})^* \rightarrow {_9F^{20}} + {_2He^4}$

followed by $_9F^{20} \rightarrow {_{10}Ne^{20}} + \beta^{-1}$, $T = 10.7$ sec.

$_{13}Al^{27} + {_0n^1} \rightarrow (_{13}Al^{28})^* \rightarrow {_{11}Na^{24}} + {_2He^4}$

followed by $_{11}Na^{24} \rightarrow {_{12}Mg^{24}} + \beta^-$, $T = 15.0$ hr.

(c) **(n-d) and (n-t) Reactions :** (n, d) reactions have been detected as a result of the bombardment of elements by 90 MeV neutrons. These reactions do not produce any compound nucleus and are direct reactions of the pick-up type. (n, t) reactions can take place in a similar manner. Tritium found in atmosphere is most likely being produced by the following reaction :

$_7N^{14} + {_0n^1} \rightarrow {_6C^{12}} + {_1H^3}$.

(d) **(n-λ) Reaction :** The radioactive capture reaction with slow neutrons is probably the most common nuclear process. It takes place with nearly all elements. The product is an isotope of the target nuclide with mass number larger by unity. Few examples are :

$_1H^1 + {_0n^1} \rightarrow (_1H^2)^* \rightarrow {_1H^2} + \gamma$

$_1H^2 + {_0n^1} \rightarrow (_1H^3)^* \rightarrow {_1H^3} + \gamma$.

The capture of a neutron increases the neutron to proton ratio. If this ratio in the target nucleus is already near the upper limit of stability, the product nucleus is radioactive usually emitting β-particle. Few examples are :

$_{45}Rh^{103} + {_0n^1} \rightarrow (_{45}Rh^{104})^* \rightarrow {_{45}Rh^{104}}$ (42 sec) + γ

$_{49}In^{115} + {_0n^1} \rightarrow (_{49}In^{116})^* \rightarrow {_{49}In^{116}}$ (54 min + γ

$_{92}U^{238} + {_0n^1} \rightarrow (_{92}U^{239})^* \rightarrow {_{92}U^{239}}$ (23.5 min) + γ

$$_{29}Cu^{65} + {}_0n^1 \rightarrow ({}_{29}Cu^{66})^* \rightarrow {}_{29}Cu^{66} \text{ (5.1 min)} + \gamma$$

$$_{79}Au^{197} + {}_0n^1 \rightarrow ({}_{79}Au^{198})^* \rightarrow {}_{79}Au^{198} \text{ (2.7 days)} + \gamma.$$

(e) **Neutron—Three or More Particles :** If the energy of the incident neutron approaches 30 MeV, sufficient energy available to overcome the electrostatic potential barrier, three neutrons or even two neutrons and a proton are ejected from the compound nucleus. When neutrons (~ 100 MeV) are used as projectiles, nuclei of moderate mass number may undergo *spallation* and those of high mass number, *e.g.*, Bi and *Pb suffer fission* probably accompanied by spallation.

(f) **(e-2n) Reaction :** In this type of reaction, one neutron is captured by the nucleus and two neutrons are emitted leaving the charge of the nucleus unchanged and less the mass number by one unit. Thus the residual nucleus is the isotope of the target nucleus. Q-value of the reaction is negative and fast neutrons are needed to bring about this reaction. In most of the cases residual nucleus is unstable and followed by position emission. Few examples are:

$$_6C^{12} + {}_0n^1 \rightarrow ({}_6C^{13})^* \rightarrow {}_6C^{11} \text{ (20.4 min)} + 2\ {}_0n^1$$

$$_{19}K^{39} + {}_0n^1 \rightarrow ({}_{19}K^{40})^* \rightarrow {}_{19}K^{38} \text{ (7.5 min)} + 2\ {}_0n^1$$

$$_{51}Sb^{121} + {}_0n^1 \rightarrow ({}_{51}Sb^{122})^* \rightarrow {}_{51}Sb^{120} \text{ (16.6 min)} + 2\ {}_0n^1.$$

$_{80}Hg^{197}$, obtained in the $_{80}Hg^{198}$ (n, 2n)$_{80}Hg^{197}$ process, decays by equivalent orbital electron capture. The emission of an electron occurs in a number of instances, especially where neutron to proton ratio is at the upper limit of the stability for the given atomic number. Original target nuclei of this type are Ge^{76}, Cd^{116}, Ce^{142} and U^{238}.

4. Transmutation by Deuterons

A great many nuclear reactions have been observed with high energy deuterons as the bombarding particles, because about 2 MeV is sufficient to cause its rupture into a neutron and proton.

(a) **(d-p) Reaction :** Reactions of this type are quite common, and the product is always an isotope of the target element with mass number one unit higher. Few examples are :

$$_6C^{12} + {}_1H^2 \rightarrow ({}_7N^{14})^* \rightarrow {}_6C^{13} + {}_1H^1$$

$_3Li^7 + {}_1H^2 \rightarrow ({}_4Be^9)^* \rightarrow {}_3Li^8 + {}_1H^1$ (electron-decay)

$_{11}Na^{23} + {}_1H^2 \rightarrow ({}_{12}Mg^{25})^* \rightarrow {}_{11}Na^{24} + {}_1H^1$ (electron-decay)

$_{15}P^{31} + {}_1H^2 \rightarrow ({}_{16}S^{33})^* \rightarrow {}_{15}P^{32} + {}_1H^1$

$_{48}Cd^{114} + {}_1H^2 \rightarrow ({}_{49}In^{116})^* \rightarrow {}_{48}Cd^{115} + {}_1H^1$

$_{83}Bi^{209} + {}_1H^2 \rightarrow ({}_{84}Po^{211})^* \rightarrow {}_{83}Bi^{210} + {}_1H^1$.

(b) **(d-α) Reaction :** Because of the high nuclear potential energy barrier for α-particles, which increases with Z, reactions of this type are observed only with high energy deuterons and elements of fairly low Z. Few examples are:

$_3Li^6 + {}_1H^2 \rightarrow ({}_4Be^8)^* \rightarrow {}_2He^4 + {}_2He^4$

$_8O^{16} + {}_1H^2 \rightarrow ({}_9F^{18})^* \rightarrow {}_7N^{14} + {}_2He^4$

$_{10}Ne^{20} + {}_1H^2 \rightarrow ({}_{11}Na^{2q})^* \rightarrow {}_9F^{18} + {}_2He^4$

$_{12}Mg^{26} + {}_1H^2 \rightarrow ({}_{13}Al^{28})^* \rightarrow {}_{11}Na^{24} + {}_2He^4$

$_{13}Al^{27} + {}_1H^2 \rightarrow ({}_{14}Si^{29})^* \rightarrow {}_{12}Mg^{25} + {}_2He^4$.

(c) The probability of radiative capture reaction is quite small because the extra nucleons in a compound are in virtual states. When the energy of the incident deuterons is about 20 MeV or more, reactions of the types (d, 2n), (d, 3n), (d, 2p) and (d, pα) are possible.

(d) **(d-t) Reaction :** Although the probability of the reaction of this type is low, some processes are known. Among these are:

$_3Li^7 + {}_1H^2 \rightarrow ({}_4Be^9)^* \rightarrow {}_3Li^6 + {}_1H^3$

$_{15}P^{31} + {}_1H^2 \rightarrow ({}_{16}S^{83})^* \rightarrow {}_{15}P^{30} + {}_1H^3$

$_{47}Ag^{109} + {}_1H^2 \rightarrow ({}_{48}Cd^{111})^* \rightarrow {}_{47}Ag^{108} + {}_1H^3$

$_4Be^9 + {}_1H^2 \rightarrow ({}_5B^{11})^* \rightarrow {}_4Be^8 + {}_1H^3$.

(e) **(d-n) Reaction :** A large number of transmutations of this

$_3Li^7 + {}_1H^2 \rightarrow ({}_4Be^9)^* \rightarrow {}_4Be^8 + {}_0n^1$

$_4Be^9 + {}_1H^2 \rightarrow ({}_5B^{11})^* \rightarrow {}_5B^{10} + {}_0n^1$

$_6C^{12} + {}_1H^2 \rightarrow ({}_7N^{14})^* \rightarrow {}_7N^{13} + {}_0n^1$ (position decay)

When two deuterons interact both the (d, n) and the (d, p) reactions have been observed.

$$_1H^2 + {}_1H^2 \rightarrow ({}_2He^4)^* \rightarrow {}_1H^3 + {}_1H^1 + 4.02 \text{ MeV}$$

$$\rightarrow {}_2He^3 + {}_0n^1 + 3.25 \text{ MeV}$$

The tritium is unstable and has a half life of about 12 years.

At high energies (> 50 MeV), the (d, n) reaction becomes a stripping process that does not involve compound nucleus formation.

5. Transmutation of Particles of Mass Three

Some reactions obtained with tritium and helium 3 are given below :

$$_3Li^6 + {}_1H^3 \rightarrow ({}_4Be^9)^* \rightarrow {}_3Li^7 + {}_1H^2$$

$$_{27}Co^{59} + {}_1H^3 \rightarrow ({}_{28}Ni^{62})^* \rightarrow {}_{27}Co^{60} + {}_1H^2$$

$$_{29}Cu^{63} + {}_1H^3 \rightarrow ({}_{30}Zn^{66})^* \rightarrow {}_{29}Cu^{64} + {}_1H^2$$

$$_1H^2 + {}_1H^3 \rightarrow ({}_2He^5)^* \rightarrow {}_2He^4 + {}_0n^1$$

$$_{16}S^{32} + {}_1H^3 \rightarrow ({}_{17}Cl^{35})^* \rightarrow {}_{17}Cl^{34} + {}_0n^1$$

$$_{13}Al^{27} + {}_1H^3 \rightarrow ({}_{14}Si^{30})^* \rightarrow {}_{12}Mg^{27} + {}_2He^3$$

$$_3Li^7 + {}_1H^3 \rightarrow ({}_4Be^{10})^* \rightarrow {}_2He^6 + {}_2He^4$$

$$_1H^2 + {}_2He^3 \rightarrow ({}_3Li^5)^* \rightarrow {}_2He^4 + {}_1H^1$$

$$_1H^3 + {}_2He^3 \rightarrow ({}_3Li^6)^* \rightarrow {}_2He^5 + {}_1H^1$$

$$_2He^3 + {}_2He^3 \rightarrow ({}_4Be^6)^* \rightarrow {}_3Li^5 + {}_1H^1$$

$$_7N^{14} + {}_2He^3 \rightarrow ({}_9F^{17})^* \rightarrow {}_9F^{16} + {}_0n^1$$

$$_6C^{12} + {}_2He^3 \rightarrow ({}_8O^{15})^* \rightarrow {}_7N^{13} + {}_1H^2$$

$$_6C^{12} + {}_2He^3 \rightarrow ({}_8O^{16})^* \rightarrow {}_7N^{13} + {}_1H^3$$

$$_7N^{14} + {}_2He^3 \rightarrow ({}_9F^{17})^* \rightarrow {}_7N^{13} + {}_2He^4.$$

6. Transmutation by Radiation

Reactions of this type brought about by high energy radiations, have been referred to as photo disintegrations. To produce this process the mean of the incident photon must exceed the binding energy of the particle to be ejected. If the energy of the γ-radiation is not sufficient to remove even neutron, the nucleus may enter one of its excited states. The excess energy is then emitted as radiation. Examples are :

$$_1H^2 + \gamma \rightarrow ({}_1H^2)^* \rightarrow {}_1H^1 + {}_0n^1$$

$$_4Be^9 + \gamma \rightarrow (_4Be^9)^* \rightarrow {}_4Be^8 + {}_0n^1$$
$$_{15}P^{31} + \gamma \rightarrow (_{15}P^{31})^* \rightarrow {}_{15}P^{30} + {}_0n^1$$
$$_4Be^9 + \gamma \rightarrow (_4Be^9)^* \rightarrow {}_3Li^8 + {}_1H^1$$
$$_{12}Mg^{25} + \gamma \rightarrow (_{12}Mg^{25})^* \rightarrow {}_{11}Na^{24} + {}_1H^1$$
$$_5B^{10} + \gamma \rightarrow (_5B^{10})^* \rightarrow {}_4Be^8 + {}_1H^2$$
$$_5B^{11} + \gamma \rightarrow (_5B^{11})^* \rightarrow {}_4Be^8 + {}_1H^3$$
$$_6C^{12} + \gamma \rightarrow (_6C^{12})^* \rightarrow {}_4Be^8 + {}_2He^4$$
$$_{36}Kr^{83} + \gamma \rightarrow (_{36}Kr^{83})^* \rightarrow {}_{36}Kr^{83} + \gamma'$$
$$_{38}Sr^{87} + \gamma \rightarrow (_{38}Sr87)^* \rightarrow {}_{38}Sr^{87} + \gamma'.$$

One nucleon either neutron or proton can be emitted when γ-rays are of energy of the order of 10 MeV. At high energies reactions of the types (γ, np), (γ, 2n), (γ, n 2p) and others, involving the ejection of two or three or more particles, have been observed. The examples are :

$$_8O^{16} + \gamma \rightarrow (_8O^{16})^* \rightarrow {}_6C^{11} + {}_0n^1 + {}_0n^1 + {}_0n^1 + {}_1H^1 + {}_1H^1$$
$$_6C^{12} + \gamma \rightarrow (_6C^{12})^* \rightarrow {}_2He^4 + {}_2He^4 + {}_2He^4.$$

7. Nuclear Reactions with Heavy Ions

Nuclear reactions with so-called heavy ions are those in which particles heavier than α-particles are used as projectiles. These are divided as :

(a) *Fission spallation and fragmentation reactions,* especially of the heavier elements are included in the third category. Mention may also be made of another kind of reaction which has been observed with target nuclei of moderate mass number, *e.g.*, the products of the interaction of 125 MeV nitrogen ions with aluminium 27 range from C^{12} to K^{38}.

$$_{13}Al^{27} + {}_7N^{14} \rightarrow (_{20}Ca^{41})^* \rightarrow {}_1H^1 + 2{}_0n^1 + {}_{19}K^{38}.$$

Similar reactions, perhaps involving the emission of alpha particles, can lead to $_{14}Si^{31}$ and other nuclei in the same mass region.

(b) The first type, called *transfer reaction,* involves the transfer of a nucleon or a few nucleons from one nucleus to the other. The examples are :

$$_7N^{14} + {_5B^{10}} \rightarrow {_7N^{13}} + {_5B^{11}}$$

$$\rightarrow {_7N^{13}} + {_7N^{15}}$$

$$_7N^{14} + {_7N^{14}}$$

$$\rightarrow {_6C^{13}} + {_8O^{15}}$$

$$\rightarrow {_{13}Al^{25}} + {_7N^{16}}$$

$$_{13}Al^{27} + {_7N^{14}}$$

$$\rightarrow {_{12}Mg^{27}} + {_8O^{14}}.$$

These reactions are known as *direct reactions* as they do not form any compound nucleus.

(c) In the second class of reactions with accelerated heavy ions, the compound nucleus is formed and two, three, four or more particles, mostly neutrons sometimes accompanied by one or two protons or α-particles are evaporated from the compound nucleus. The higher the energy of the projectile, the larger the number of particles expelled. The examples are :

Nb^{93} (C^{12}, 2n) Ag^{103}, Au^{197} (C^{12}, 4n) At^{205}, Al^{27} (C^{12}, αn) Cl^{34}, U^{238} (N^{14}, α5n) Bk^{243} and V^{51} (O^{16}, 2n) Ga^{65}.

CONTINUUM THEORY OF NUCLEAR REACTION

At higher bombarding energies the individual levels of the compound nucleus become broader and also more closely spaced. The total width Γ becomes much greater than D, the spacing of the levels. The width and number of levels are such that even though the interval is small, many levels may be found in it. Sharp resonances are no longer observable when $\Gamma \geq D$. The spacing between levels is completely occupied, so the *space is described as continuum.* The cross-section for the formation of a compound nucleus is larger for neutrons than for charged particles because Coulomb repulsion between the incident charged particle and the target nucleus is important in the latter case. In the medium and heavy nuclei, the individual level becomes broader and levels become more closer when the energy of the incident particle is large. The continuum theory of nuclear reaction cross-sections treats the individual level not separately but as an average over many resonances.

We shall derive the value of cross-section on the basis of classical mechanics. If the incident particle is a neutron (with J = 0), the cross-section for reaching the nuclear surface is given by the classical target area such as,

$$\sigma_C(n) = \pi R^2. \qquad ...(1)$$

We now proceed to a qualitative wave mechanical discussion. This discussion is based upon the fact that there is a sudden change of potential when the particle crosses the boundary of the nucleus, due to the strongly attractive nuclear forces. The sudden change of potential reflects the incoming wave at the nuclear surface and thus decreases the cross-section, especially at low energy.

Let us consider a beam of particles of energy E moving in the direction of x in a potential V(x), which is zero for $x < 0$ and is $-V_0$ for $x > 0$. The particles are partially reflected at $x = 0$. We have to determine now the transmission coefficient T, defined as the ratio of the number of particles transmitted through the region $x > 0$ to the number of incident particles. Thus

$$\text{For } x < 0,\ \psi = Ae^{ikx} + Be^{ikx}, \qquad ...(2)$$

where
$$k = \sqrt{\frac{(2ME)}{\hbar}}. \qquad ...(3)$$

$$\text{For } x > 0,\ \psi = Ce^{+iKx} \qquad ...(4)$$

where
$$K = \sqrt{\frac{[2M(E+V_0)]}{\hbar}} \qquad ...(5)$$

Here A, B and C are coefficients giving the amplitude of the various wave functions. Since neither the wave function nor the kinetic energy can become infinite at any point, hence both, the wave function and its derivative must be continuous at all points. Using boundary conditions at $x = 0$, we have

$$A + B = C \text{ and } ikA - ikB = iKC.$$

$$\therefore\ A = \frac{1}{2}\left(\frac{1+K}{k}\right) \text{ and } B = \frac{1}{2}C\left(\frac{1-K}{k}\right) \qquad ...(6)$$

$$\text{Hence } T = \frac{|A|^2 |B|^2}{|A|^2} = \frac{4Kk}{(K+k)^2} \qquad ...(7)$$

Hence limiting values of T are of special interest. At high incident energies relation (7) indicates that K = k and thus T approaches 1. On the other hand, as the neutron energy approaches zero,

$$k < K \text{ and } T = \simeq \frac{4k}{K} \simeq 4\left(\frac{E}{V_0}\right)^{1/2}.$$

The position of the particle is uncertain by the amount of its wavelength $\lambdabar$. This can be accounted approximately for by replacing πR^2 with $\pi(R + \lambdabar)^2$. Thus cross-section becomes

$$\sigma_c = \pi(R + \lambdabar)^2. \qquad ...(8)$$

This equation has to be modified by including the transparency T of the potential barrier. Thus we obtain the cross-section for the formulation of the compound nucleus by S-wave neutrons.

$$\sigma_c = \pi(R + \lambdabar)^2 4kK/(k + K)^2. \qquad ...(9)$$

For very small neutron energies $k \leq K$ and $\lambdabar \geq R$, the above equation becomes

$$\sigma_c = \frac{4\pi}{kK}. \left(\text{Because } k = \frac{1}{\lambdabar}\right) \qquad ...(10)$$

At high energies the cross-section approaches the classical value πR^2 and increases with decreasing energy, but less strongly than $\pi(R + \lambdabar)^2$, because of the reflection at the nuclear boundary.

Similar approach can also be used to obtain an expression for the reaction cross-section of charged particles. In this case the incident beam is deviated by the Coulomb potential $V(r) = Z_1 Z_2 e^2/2\pi\epsilon_0 r$. The incident particles reaching the nuclear surface thus have a maximum impact parameter $R[1 - (B/E)]^{1/2}$, where B = Coulomb barrier height = V(R). Hence the reaction cross-section can be expressed as

$$\sigma_c = \pi R^2[1 - B/E] \qquad E > B$$

$$= 0 \quad E \leq B. \qquad ...(11)$$

Here we see that the high energy limit of this eqn. is πR^2, identical to that for neutrons.

An attempt to obtain a better fit to the continuum theory value of σ_c has been made by Destrovsky *et al.*, in 1959.

RESONANCE : BREIT WIGNER DISPERSION FORMULA

The concept of cross-section and level width can be applied to resonances in a quantitative way. In the particularly important case of resonance processes, a theoretical formula for the cross-section was derived by G. Breit and E.P., Wigner in the United States in 1936. In its simplest form, it gives the value of the cross-section in the neighbourhood of a single resonance level formed by an incident particle with zero angular momentum and charge zero so that spin and the Coulomb effects can be ignored. The result is analogous to the theory of optical dispersion, so that the main formula obtained is often called the *dispersion formula.*

Whether or not the level of compound nucleus is bound, excitation by an incident particle can be treated as analogous to the excitation of the oscillations produced in an electric circuit by an electromagnetic wave. We therefore, expect the nuclear cross-section to vary with the incident energy in the same way that the energy in a forced oscillation varies with incident frequency. The classical resonant circuit absorbs energy because of resistive levels. In the nuclear case, damping arises because of the possibility of decay. Because of this possibility the nuclear state has a finite width Γ. The wave function of a decaying state of mean energy E_0 may be written as

$$\psi(r,t) = \psi(r)e^{iE_0 t/\hbar}e^{-\Gamma t/2\hbar}. \qquad ...(1)$$

This corresponds to an exponential decrease of intensity of excitation $|\psi(r, t)|^2$ with a time constant $\tau(=\hbar/\Gamma)$. This wave function also shows that a decaying state is not a state of definite energy E of the form $\Psi(r)e^{-i(E/\hbar)t}$. Never the less, it can be represented by superposition of states of slightly different energies E, each with a different amplitude A(E)

$$\Psi(r, t) = \int_{-\infty}^{+\infty} A(E)e^{-iEt/\hbar}dE. \qquad ...(2)$$

Using the Fourier analysis technique, we can show that the energies E are grouped about a mean energy E_0 with a spread of the order of $\Gamma = \hbar\lambda$. Equating eqns. (1) and (2), we get

$$e^{-\Gamma t/2\hbar} = e^{-\lambda t/2} = \int_{-\infty}^{+\infty} A(E)e^{-iEt/\hbar}dE. \qquad ...(3)$$

According to the Fourier theorem any well behaved function f(t) can be represented as

$$f(t)=\frac{1}{2\pi}\lim_{s\to\delta}\int_{-\Omega}^{\Omega}-i\omega t\,d\omega\int_{-\infty}^{\infty}e^{j\omega t'}f(t')dt'. \quad ...(4)$$

Applying this to the function $e^{-\lambda_t/2}$, we get

$$4(E)=\frac{1}{2\pi\hbar}\int_0^{\infty}e^{\left[i(E-E_0)/\hbar-\lambda/2\right]t'}dt'$$

$$=\frac{1}{2\pi}\frac{1}{(E-E_0)+i\hbar\lambda/2}\cdot \quad ...(5)$$

Here we have assumed that the decaying system was prepared at the time t = 0. The probability of finding the system with a given energy E is proportional to

$$|A(E)|^2=\frac{1}{4\pi^2}\frac{1}{(E-E_0)^2+(\hbar\lambda/2)^2}$$

$$\frac{1}{4}=\frac{1}{4\pi^2}\frac{1}{(E-E_0)^2+\frac{\Gamma^2}{4}}\cdot \quad ...(6)$$

This gives the level shape. It is exactly as for pure radiative decay except that particle emission is now included by using the total width Γ instead of the radiative width Γ_r. The cross-section for excitation of the level by collision of particle x width nucleus X is, therefore, expected to have the form

$$\sigma_x=\frac{C}{\left[(E-E_0)^2+\frac{\Gamma^2}{4}\right]}, \quad ...(7)$$

where C is constant. For the determination of C, let us suppose that the compound nucleus formation and decay processes take place in a box of volume V which contains one nucleus X and one particle x. If the states are quantised, the no. of states of motion of particles with momentum between p and p + dp = $(4\pi p^2 dpV)/h^3$. The probability of formation of the compound level per unit time = No. of possible states of motion × probability that the nucleus X is contained within the small volume σ_{xv} out by the

$$dP=\frac{4\pi p^2dp}{h^3}V\cdot\frac{\sigma_{xv}}{V}=\frac{4\pi}{h^3}=v\sigma_x p^2dp.$$

Substituting the value of σ_x from eqn. (7) in the above eqn. and integrating over the energy spectrum we get

$$P = \int_{-\infty}^{+\infty} \frac{4\pi}{h^3} v \cdot p^2 \cdot \frac{C}{(E-E_0)^2 + \frac{\Gamma^2}{4}} dt$$

$$= \int_{-\infty}^{+\infty} \frac{4\pi}{h^3} \cdot \frac{1}{(E-E_0)^2 + \frac{\Gamma^2}{4}} dE \; .$$

If we assume that the variation of the channel wavelength λ of the particle over the level width Γ may be neglected, then the probability of formation per sec

$$= \frac{4\pi C}{h\lambda^2} \cdot \frac{2}{\Gamma} \left[\tan^{-1} \frac{2(E-E_0)}{\Gamma} \right]_{-\infty}^{+\infty}$$

$$= \frac{4\pi C}{h\lambda^2} \cdot \frac{2}{\Gamma} \cdot \pi = \frac{C}{\hbar \pi \lambda\!\!\!^{-2} \Gamma} . \quad \text{...(8)}$$

Probability of decay per sec $= \dfrac{\Gamma_x}{\hbar}$,

where Γ_x is the partial width of the compound level for the emission of x. In the equilibrium state the probability of formation per sec would equal to the probability of decay per sec of the excited state back into the system X + x. Hence

$$\frac{C}{\pi \hbar \lambda\!\!\!^{-2} \Gamma} = \frac{\Gamma_x}{\hbar} \text{ or } C = \pi \lambda\!\!\!^{-} \Gamma \Gamma_x, \quad \text{...(9)}$$

Substituting this value in eqn. (7), the cross-section for the formation of the level becomes

$$\sigma_x(E) = \frac{\pi \lambda\!\!\!^{-2} \Gamma \Gamma_x}{(E-E_0)^2 + \frac{\Gamma^2}{4}} \quad \text{...(10)}$$

For the process X(x, y) Y, we obtain reaction cross-section as

$$\sigma_r = \sigma_{xy}(E) = \frac{\sigma_x \Gamma_y}{\Gamma} = \pi \lambda\!\!\!^{-2} \frac{\Gamma_x \Gamma_y}{(E-E_0)^2 + \frac{\Gamma^2}{4}} . \quad \text{...(11)}$$

Here Γ_x and Γ_y are partial level width defined as

$$\Gamma_{x\tau x} = \hbar \text{ and } \Gamma_{v\tau v} = \hbar$$

where τ_x and τ_v the main life times that the compound nucleus would have if elastic scattering of x or the emission of y were the only possible modes of decay.

If spin is considered, the right band side of eqn. (11) must be multiplied by the factor

$$gc = \frac{(2Ic+1)}{(2I_x+1)(2Ix+1)}, \quad \text{...(12)}$$

where I_x is the total angular momentum of the incident particle, Ix is that of target nucleus and Ic is of the compound state, which is formed only by those orbital angular momentum l_x which satisfy the conditions

$$Ic + Ix + I_x + I_x \text{ and } \Pi x \Pi_x (-1)^{l x} = \Pi c. \quad \text{...(13)}$$

Hence for the nuclear reaction in which particles have definite spins, the relation (7) can be written as

$$\sigma_r = \pi \lambda^2 \frac{(2Ic+1)}{(2I_x+1)(2Ix+1)} \frac{\Gamma_x \Gamma_y}{(E-E_0)^2 + \frac{\Gamma_v^2}{4}} \quad \text{...(14)}$$

This is known as the *Breit Winger resonance formula.* For the (n, γ) reaction in particular

$$\sigma(n, \gamma) = \pi \lambda^2 = \frac{(2Ic+1)}{2(2Ix+1)} \frac{\Gamma_x \Gamma_y}{(E-E_0)^2 + \frac{\Gamma^2}{4}} \quad \text{...(15)}$$

This is maximum when $E = E_0$ and is equal to

$$\sigma_{max} (n, \gamma) = 4\pi \lambda^2 gc \Gamma_n \Gamma_\gamma / \Gamma^2. \quad \text{...(16)}$$

For $E = E_0 \pm \Gamma/2$, $\sigma = 1/2\ \sigma_{max}$ and hence Γ is the full width at half maximum. A sharp resonance corresponds to a narrow width Γ and thus to an excited state of long life. A largest possible capture cross-section will occur when $\Gamma_n = \Gamma_\gamma = \Gamma/2$. Its maximum possible value σ_{max} = $\pi gc \lambda'^2$. The width of the resonance peak affects the cross-section. In a general way, if the peak is broad, covering a large energy range, the cross-sections are likely to be somewhat decreased as compared with the case of sharp and narrow peak. The width of the peak is inversely related with the life of the excited state of the compound nucleus. The total cross-section σ_t and the radioactive capture cross-section σ(n, γ) have been measured as functions of the incident neutron energy. Some of the results are shown in Fig. 2.12 in which these cross-sections are plotted as functions of energy in the intermediate range (10 – 500 keV). In this energy range total cross-section for Al^{27} is practically equal to the elastic scattering cross-section as only reactions in this case are the elastic scattering and radioactive capture.

Equation (14) applies to all resonance cross-sections except for the elastic re-emission of the incident particle σ(x, x), *i.e., elastic resonance scattering.* Besides compound nucleus scattering (*i.e.,* absorption and re-emission of neutron of the same energy), the incident particle wave is scattered by the nucleus as if it were an impenetrable sphere. This type of scattering is known as *potential* or *shape-elastic scattering.* The elastic-scattering cross-section σ_{el} is given by

$$\sigma_{el} = 4\pi\lambda\!\!\!^{-2}\left[gc\left| \frac{\Gamma_{n/2}}{\left(E - E_0 + \dfrac{i\Gamma}{2}\right)} + e^{i\phi_l}\sin\phi_l \right|^2 + (1-gc)\sin^2 f_l \right], \quad ...(17)$$

where ϕ_l is an energy dependence quantity known as the hard sphere phase shift. For $l = 0$ or S-wave neutrons $\phi = R/\lambda\!\!\!^- = Rk$. Hence for S-wave neutrons on a spinless target

$$\sigma_{el} = 4\pi\lambda\!\!\!^{-2}\left| \frac{\Gamma_{n/2}}{(E - E_0) + i(\Gamma/2)} + e^{kR}\sin kR \right|^2 \quad ...(18)$$

Here the coefficient $4\pi\lambda\!\!\!^{-2}$ is the maximum possible S-wave scattering cross-section. The first term between bars is called the resonance scattering amplitude, which if present alone would lead to eqn. (9). The second term is called the *potential scattering amplitude,* which if present alone would lead to

$$\sigma_{el} = 4\pi\lambda\!\!\!^{-2}\sin^2 kR. \quad ...(19)$$

This term varies smoothly with energy. For small energies $kR \leq 1$. As $E \to 0$, we get

$$\sigma_{el} \simeq 4\pi R^2, \quad ...(20)$$

which is the potential scattering cross-section for an impenetrable sphere when $R \leq \lambda\!\!\!^-$.

The resonance term rises to large values near $E = E_0$, but is small elsewhere. For $E < E_0$ the two terms interfere destructively yielding a low value of σ_{el}. For $E > E_0$ the interference is constructive. For charged particles, resonance scattering involves coherence between Coulomb potential scattering, nuclear potential scattering and resonance scattering. Let us consider the lowest energy resonance of the compound nucleus. For $E \leq E_0$, the denominator of eqn. (15) does not change much with E and Γ_γ is independent of E but Γ_n does depend on E. Hence $\sigma(n, \gamma) \propto \lambda\!\!\!^-{}^2 \, \Gamma_n$. The probability of elastic emission of a neutron of energy E (momentum p) is proportional to the density of states in momentum space around p. Hence

$$\Gamma_n \propto \frac{p^2 dp}{dE} \alpha p \begin{Bmatrix} E = \dfrac{p^2}{2m} \\ dE = \left(\dfrac{p}{m}\right) dp \end{Bmatrix}$$

Thus we have $\sigma(n, \gamma) \propto \lambda\!\!\!^-{}^2 p$

$$\propto \frac{1}{p} \qquad \left(\because \lambda\!\!\!^- = \frac{\hbar}{p}\right)$$

or $$\propto \frac{1}{v}, \qquad ...(21)$$

where v is the velocity of neutron. *It is known as l/v – law.* Thus Breit-Wigner formula leads to the conclusion that at low neutron energies the cross-section should be inversely proportional to the neutron speed. Following 1/v *region* there occurs the *resonance region* in which the cross-sections rise sharply to high values. The cross-sections are low with neutrons of very high energy.

OPTICAL MODEL

The continuum theory does not stand upto experimental tests satisfactorily. Barschall plotted the measured total neutron cross-sections in a three dimensional graph against the neutron energy E and the mass number A, and found that the cross-sections did not decrease smoothly with increasing E, as predicted by the theory, and that the trend of those maxima and minima with energy was a smooth function of A. He showed that the disagreement with the theory was not due to unexpected resonances in individual nuclides, but to a general flow in our theory. The widely spaced shallow maxima and minima were expected from scattering by a potential well, as in the shell model. Thus we see that *our compound nucleus theory must be modified in the light of shell model idea.*

A mathematical mode, known as *optical model,* pictures the interaction among the nucleus as being intermediate to that predicted by the continuum and shell models of the nucleus. It is the model of a complex nuclear potential, the real part produces a potential scattering like the scattering by a hard sphere and imaginary part corresponds to the cross-section for compound nucleus formation. The nucleus can thus be viewed as a *cloudy crystal ball,* which is able to transmit, refract and absorb the incident particle waves, analogous to the transmission, reflection, refraction and absorption of a light beam. In optics we require an index of refraction which is a complex quantity to describe optical phenomena. Due to this analogy with optics, this nuclear model is named as *optical model.*

Let us first discuss the optical model of Fernbach, Serber and Taylor, which involves a nuclear potential of complex form. For the sake of simplicity, we consider the case that the real and imaginary parts of the nuclear potential have the same radial dependence, such as

$$V = V(r)\,(1 + i\xi) \quad \text{for } r \le R$$
$$V = 0 \qquad \text{for } r \ge R. \qquad ...(1)$$

The Schrodinger equation of the neutron with this potential is

$$\nabla^2\psi + \frac{2M}{\hbar^2}\,[E - V(r)\,(1 + i\xi)]\,\psi = 0. \qquad ...(2)$$

For the special case of $l = 0$ neutrons and a square well of depth V_0, for which

$$V(r) = -V_0\ (r < R),\ V(r) = 0\ (r > R),$$

the radial equation is given by

$$\frac{du^2}{dr^2} + K_c^2 u = 0, \quad r < R \qquad ...(3)$$

and $$\frac{d^2u}{dr^2} + k^2 u = 0 \quad r > R, \qquad ...(4)$$

where

$$Kc = \frac{\sqrt{[2M(E + V_0 + i\xi V_0)]}}{\hbar} = \left(\frac{2M}{\hbar^2}\right)^{1/2}(E + V_0)^{1/2}\left(1 + \frac{i\xi V_0}{E + V_0}\right)^{1/2}$$

$$= \sqrt{\frac{[2M(E+V_0)]}{\hbar + i\xi V_0\left[\dfrac{M}{2\hbar^2(E+V_0)}\right]^{1/2}}} = K + ik'$$

Hence the solution of eqn. (110) is of the form

$$\psi(r) = \left(\frac{A}{r}\right)\sin Kcr$$

or $$|r\psi|^2 = A^2[\sin^2 Kr + \sinh^2 k'r), \text{ for } r \le R. \quad ...(5)$$

The wave is absorbed as it penetrates into the potential step. The characteristic length known as penetration depth for $E \le V_0$ is given by

$$L = \frac{1}{k'} = \frac{2\hbar(E+V_0)^{1/2}}{\xi V_0\sqrt{(2M)}} = \frac{2\hbar}{\xi(2MV_0)^{1/2}} = \frac{2}{\xi K} \quad ...(6)$$

Substitution of experimental values ($x \approx 0.06$, $V_0 \approx 50$ MeV), we get L of the order of nuclear diameter. This model has explained discontinuities in the variation of the elastic cross-sections for slow neutrons as a function of the mass number of nuclei. The maxima has been observed for radii $R = \left(n + \frac{1}{2}\right)\lambda_{inside}$. The study of elastic scattering has provided valuable information on nuclei radii. The radii so calculated were not corresponding to the relation $R = R_0A^{1/3}$. The above potential form gives too large a cross-section for the elastic scattering process in backward directions.

A much better fit to the experimental cross-sections is the potential well proposed by Woods and Saxon. In it V(r) varies smoothly with r, the distance from the centre of the nucleus, and is of the form

$$V(r) = -(V_0 + iW_0)\,[1 + e^{(r-R)/a}] \quad ...(7)$$

The plot of its real part is shown in Fig. 2.14. The curve shows that the potential well is no longer square but has rounded edges. R is the radius of the equivalent square well and is very nearly equal to the radius at which the depth of the potential well is one-half its central value a is a surface density parameter and is a measure of the degree of the tapering of the potential near the nuclear surface. The optical model parameter V_0 and W_0 (= ξV_0) are dependent on the type and energy of the incident particle, V_0 decreases smoothly with the incident energy.

This model gives little idea about the nuclear surface. A better fit to the experimental results is obtained if a surface absorption potential of Gaussian shape, centred around r = R is included.

In recent years other forms of optical potentials have been introduced to take into account the deformation of the nuclei, the spin-orbit coupling and also the energy region at which the experiments are being performed. Additional terms are required to take into account these effects. *The real part the potential function accounts for refraction and imaginary part for absorption of the incident nucleons.*

The spin-orbit term describes changes in spin orientations through the coupling of spin and angular momenta. This orientation is known as polarization. In addition to the theory of the neutron-nuclear interaction, the optical model has been successively applied to the scattering of protons, deuterons, alpha particles and heavier ions. We *shall give few experimental results.* R.D.

Albert in 1959 has measured (p, n) cross-sections in some medium weight nuclei between the energy range 4 to 5.5 MeV, and used the Bjorklund and Fernback optical potential which has the form

$$U(r) = V_\rho(r) + iWq(r) + \frac{\lambda\hbar^2}{4m_0^2c^2}\,\frac{V}{r}\,\frac{d\rho}{dr}\,(\vec{\sigma}.\vec{I}) \qquad ...(8)$$

where $\rho(r) = [1 + e^{(r-R)/a}]^{-1}$, $q(r) = e^{-[(r-R)/b]^2}$, $R = r_0A^{1/3}$ and m_0 = mass of the electron.

The theoretical curves, obtained with parameters of the following table, adopted as giving the best fits to experimental data are also indicated by the thick lines. These results show that there is an excellent agreement between theory and experiment.

The optical potential function represented by equation (8) has also been used by Shore, Wall and Irvine to analyze angular distribution of 7.5 MeV protons scattered elastically by vanadium. They used two sets of parameters corresponding to surface absorption and to volume absorption. Recently, Rosen, Beery, Goldhaber and Auerbach have proposed a spin dependent optical potential by analyzing and measuring 80 separate angular distribution data.

Table 1

E(MeV)	V(MeV)	W(MeV)
4.0	47	7.0
4.5	45	7.7
5.0	45	8.25
5.5	44	8.7

Halbert, Bassel Satchler carried out an analysis of the d-d scattering data at 11 MeV and at 11.8 MeV. They used nuclear potentials of the form

$$V(r) = V\left[1 + e^{\left(r - r_0 A^{1/3}\right)/a}\right]^{-1} + iW\left[1 + e^{\left(r - r_w A^{1/3}\right)/a_w}\right]^{-1} \qquad \ldots(9)$$

for volume absorption and

$$V(r) = V\left[1 + e^{\left(r - r_0 A^{1/3}\right)/a}\right]^{-1} - ia_w W_0 \frac{d}{dr}\left[1 + e^{\left(r - r_w A^{1/3}\right)/a_w}\right]^{-1}$$

for surface absorption.

In each case uniform charge distribution of radius 1.3 $A^{1/3}$ F was assumed for the Coulomb potential. They obtained an excellent fit for the target nuclei Ni, Zr, Ag and Sn and a reasonably good fit for Ca and Ti.

The case of the deuteron is complicated because, due to polarization or stripping of the deuteron by Coulomb interaction, non-elastic channels may couple to the elastic channel and may change the optical parameters.

The optical potential may contain spin-orbit terms when the incident particle has a spin other than zero. The strength of the spin-orbit potential is adjusted to fit the measured polarization of the scattered particles. It has been found from experiments that the idea of the optical potential is not limited to the calculation of cross-sections of neutron reactions and incident energies of a few million electron volts but is applicable to the description of reactions of protons and heavier particles in the region of high energies as well.

The basic idea of the shell model is present in the optical model analysis of nuclear reactions. If the imaginary parts of the potential were absent, we should have an ordinary Schrodinger equation and corresponding stationary orbits. The shell model shows that such an approximation has considerable merit.

DIFFERENT STAGES OF A NUCLEAR REACTION

Detailed theories of nuclear reaction were patterned after the two nuclear models. The rigid drop model and the shell model. In the *compound nucleus* theory it was assumed that a nuclear projectile incident on a nucleus would interact strongly with all the nucleons in the nucleus and quickly share its energy with them. In the *reaction theory* (optical model theory), it was proposed that an incident nucleon would interact with the nucleus via the shell model potential. The success of the optical model suggests that the Bohr theory of compound nucleus is in need of modification. A more general scheme of nuclear reaction this been described by Weisskopf. According to him the nuclear reaction proceeds through three stages : the *independent particle (optical model) stage, compound nucleus stage and the final stage.* During the first stage, the interaction between the incident wave and nuclear potential may lead to partial reflection of the incident wave, called *shape elastic scattering.* The part of the wave function which enters the nucleus undergoes absorption. This process leads to the second stage. Some of the possible absorption processes are :

1. Ejection nucleons in a collisions with incident particle a direct interaction,
2. Multiple collisions of the incoming particle with several nucleons of the target nucleons,
3. The excitation of some type of collective motion such as surface vibrations of the target nucleus,
4. The formation of compound nucleus, without remembering details of the initial stage of formation.

The third stage of the reaction is the more or less rapid transition to the final state. It is concerned with the way the reaction products are produced, *i.e.,* on the second stage. Either disintegration or de-excitation of the compound nucleus will occur. Actually there is no sharp division between the different possibilities in a compound system. Two extreme cases, direct interaction and the compound nucleus, correspond to two general aspects individual and collective particle movements respectively.

STATISTICAL THEORY OF NUCLEAR REACTIONS

The theoretical treatment of the interactions between an incident particle and the nucleons in the nucleus requires the use of a nuclear model. In light nuclei the energy levels of the compound nucleus are generally well separated and the theory of the resonance reaction. In the heavy nuclei, the levels are closely spaced. If the excitation energy is sufficiently high, overlapping levels will be excited and the no. of excited levels is too great to be treated individually, and thus requires Fermi statistics. The compound nucleus is pictured then as a hot liquid or solid from which various particles evaporate. From heavy nuclei, neutrons evaporate most easily even at high energies of excitation but proton emission is also common from lighter nuclei.

In statistical thermodynamics, the entropy S(E) of the system is related to ρ(E), the density of states per unit energy interval, according to the Boltzmann formula as

$$S(E) = \log \rho\,(E). \qquad ...(1)$$

More formal derivation of eqn. (1) is also obtained by defining the thermodynamic free energy F(β) of the nucleus at temperature T = 1/β as

$$e^{-\beta F} \int \rho(E')e^{-\beta E'}dE'. \qquad ...(2)$$

We can invert the Laplace transform to obtain

$$\rho(E') = \frac{1}{2\pi i}\int e^{\beta(E'-F)}d\beta. \qquad ...(3)$$

This integral can be solved by using *steepest descent method (saddle point method)*. The integrated has a saddle point at β = β(E) given by d(βF)/dβ = E, where E is assumed as independent variable. This gives that

$$\frac{d}{dE}\left[\beta(E-F)\right] = \frac{d}{dE} = \beta = \frac{1}{T}, \qquad ...(4)$$

where $$S(E) = \beta(E-F) = \int_0^E \beta(E')dE$$

Thus integration of eqn. (3) gives

$$\rho(E) = \frac{e^S}{\sqrt{\left(-\dfrac{2\pi dE}{d\beta}\right)}}.$$

$$\therefore \quad \frac{d}{dE}\log \rho(E) = \frac{dS}{dE} - \frac{1}{2}\frac{d}{dE}\log\left(-\frac{dE}{d\beta}\right) = \frac{1}{T} - \frac{1}{2}\frac{d}{dE}\log \left(T^2 \frac{dE}{dT}\right)$$

As the last term of this eqn. is negligible, hence we get

$$\frac{d}{dE}\log \rho\,(E) \simeq \frac{d\,S(E)}{dE} = \frac{1}{T}. \qquad ...(5)$$

This is nothing but eqn. (117). Let us now find the dependence of nuclear entropy S and the energy E of the nucleus on the nuclear temperature T. If the nucleons in nucleus are assumed to be as an ideal Fermi gas, we know that the thermal heat capacity c = dE/dT is proportional to T and, therefore, $E \propto T^2$. Hence

$$E = aT^2, \qquad ...(6)$$

where a is a constant. From eqns (6) and (5), we get

$$\frac{d}{dE}\log \rho(E) = \left(\frac{a}{E}\right)^{1/2} \quad \text{or } \rho(E) \sim e^{2(aE)1/2} \qquad ...(7)$$

As approximate result obtained by Ericson is as given below.

$$\rho(E) = \frac{1}{16}\left(\frac{\pi^2}{a}\right)^{1/4} E^{-5/4}\, e^{2(aE)1/2} \qquad ...(8)$$

DIRECT REACTIONS

A direct reaction is one which proceeds without the formation of a compound nucleus. The time during which the incident particle interacts with the target nucleus is very much shorter than the life of a correspond compound nucleus. Formation of compound nucleus is more likely at low energies whereas the direct reaction mechanism will prevail at higher energies. The term direct reaction is used for a variety of nuclear processes including *inelastic nuclear collisions, stripping* and its inverse, the *pick up reaction.* We shall discuss briefly the stripping reaction and its inverse reaction.

Stripping Reactions

The term stripping is used for a type of direct reaction in which the incoming compound particle splits into two fragments, one of which absorbed by the target nucleus and the other continues more or less undisturbed. In this type of reaction target nucleus captures one or two, sometimes three nucleons from the incident particle without the formation of a compound nucleus as an intermediate stage. The remaining portion of the projectile is usually proton, neutron or deuteron. Stripping reactions of several types, *e.g.*, (d, p), (d, n), (t, p), (t, d) and (α, p) are known to occur at high particle energies with many different nuclei. In pick-up reaction incident particle removes one or two nucleons from the target nucleus. Reactions of this type are (p, α), (p, t), (p, He^3). (d, t) and (d, He^3). The theories of stripping and pick-up reactions are very similar and the angular distributions are also quite similar.

Stripping reaction was first recognized by Oppenheimer and Phillips in 1935 in analyzing the low energy (d, p) reactions. It was observed experimentally that (d, p) reactions were more frequent than (d, n) reactions. They explained the reaction by stating that the deuteron behaves as a resistively loose combination of a neutron and a proton. When it approaches a nucleus, the electrostatic repulsion of the positive charges tends to force the proton away, but the neutron is not affected. If the energy of the incident deuteron exceeds the neutron binding energy (2 MeV), the proton will be repelled and neutron will enter the target nucleus. The (d, n) reactions are of particular interest as they are used as a means for obtaining neutrons of high energy. In the high energy region the (d, p) and (d, n) reactions are equally probable.

At medium energies (10 – 20 MeV), the stripping reactions has been very useful for the determination of the energies as well as the spins and parities of excited states of nuclei. In the (d, p), say, stripping reaction the target nucleus accepts a neutron of orbital angular momentum l_n directly into one of the levels of the final nucleus. The proton proceeds in a direction determined by l_n and with an energy equal to the reaction energy Q for the formation of the level into which its partner was captured plus the kinetic energy of the deuteron. Consider a vector diagram for the (d, p) reaction. The deuteron approaches with momentum $k_d\hbar$ and the proton goes off with momentum $k_p\hbar$ at an angle θ with the direction of incident deuteron. The momentum of captured neutron may be obtained from the cosine rule applied to momentum triangle.

$$k_n^2 = k_d^2 + k_p^2 - 2k_d k_p \cos\theta . \quad ...(125)$$

A semi-classical consideration of the conservation of angular momentum suffices to give a qualitative understanding of the angular distribution of the protons after the (d, p) reaction. At the point of impact of the deuteron on the surface of the nucleus, there is a splitting into proton and neutron. The neutron subsequently penetrates the nucleus having an impact parameter r and is captured by it. The transfer of momentum $k_n\hbar$ is associated with the transfer of a quantized orbital angular momentum. Hence we have

$$(k_n \times r)\ \hbar = l_n\hbar \text{ or } k_nR \sin \beta = l_n. \quad ...(1)$$

If $l_n/k_n > R$, the nuclear radius, it is not possible for the d-n interaction to occur. This in order to occur a stripping process, we must have $l_n/k_n \le R$. Hence

$$k_d^2 + k_p^2 - 2k_dk_p \cos\theta \ge \frac{l_n^2}{R^2} \quad ...(2)$$

or

$$\cos\theta \le \frac{k_d^2 + k_p^2}{2k_dk_p}\ \frac{l^2}{2k_pk_dR^2}. \quad ...(3)$$

The first term on the R.H.S. is always ≥ 1, as $a^2 + b^2 \ge 2ab$. The second term increases quadratically with the quantum number l. There is a lower limit for the angle θ that depends on l_n. Equation (3) shows that the preferred angle of scattering θ will increase with increasing l_n. Choosing

$$E_d = 7 \text{ MeV} \qquad (\therefore k_d = 0.82 \text{ f}^{-1}),$$

$$E_p = 13 \text{ MeV} \qquad (\therefore k_p = 0.79 \text{ f}^{-1}),$$

R = 6f and applying cosine rule we get θ_{max} = 16°, 29° and 42° corresponding to l = 1, 2, 3 respectively. For each l_n values there are secondary maxima also. In general the differential cross-section is largest for $l_n = 0$ and decreases as the angular momentum transfer increases. The position of a peak or a pattern of peaks in the angular distribution of the emitted protons yields the value of l_n.

The analysis of the angular distribution of these protons provides information about angular momentum of these captured neutrons and on the spin and parity of the final nucleus. Let us consider the initial and final states of the nucleus with angular momenta I_i, I_f, the orbital angular momentum of deuteron with respect to the target nucleus l_d and the deuteron spin s_d. If in (d, p) stripping reaction, the proton escapes at an angle θ with respect to the initial direction of motion of the deuteron and neutron is captured, with orbital angular momentum l_n, then we can write

$$I_f = I_i + I_d + s_d - I_p - s_p = I_i + I_n + s_n. \quad ...(4)$$

The orbital angular momentum of the captured neutron l_n is thus restricted by the inequality

$$I_f + I_f + \frac{1}{2} \geq l_n \geq || I_i - I_f \left| -\frac{1}{2} \right| \quad ...(5)$$

Law of conservation of parity $II_f = II_i(-1)^{l_n}$ demands that initial and final states have same parity if l_n is even, and opposite parities if l_n is odd.

The quantum theory of low energy stripping reactions was first developed by Butler, using Born approximation which assumes that the plane or any particle whatever close to the nucleus is not distorted by the presence of nuclear potential. Agreement with experiment is excellent over a wide range of target nuclei and deuteron energies. Improvements have been made which lead to better agreement. They require the use of DWBA (distorted wave Born approximation). There are a number of and it is very likely that a better fit could be obtained. At very high energies the stripping phenomenon may be described by considering one of the nucleons as absorbed by a nucleus and the other as continuing along its trajectory almost unperturbed with the initial velocity of the deuteron. If proton is captured by the target nucleus, the energy spread and the angular spread of the neutron beam are

$$\Delta E = 1.5\,(E_d B_d)^{1/2}, \text{ and } \Delta^{\theta} = 1.6 \left(\frac{B_d}{E_d}\right)^{1/2} \quad ...(6)$$

where B_d is the binding energy of the deuteron and E_d its kinetic energy.

STRIPPING REACTIONS AND THE SHELL MODEL

It was pointed out first by Bethe and Butler in 1952 that stripping reactions in some cases could be used to measure the purity of shell model states. In stripping reaction the captured nucleon, say the neutron, occupies one of the available quantum states in the target nucleus and the other, say the proton, carries some information concerning this state. Conversely in pick-up reaction the picked up nucleon is removed from an occupied quantum state and the emerging deuteron conveys some information about this level. Thus we see that the stripping can be used the study unoccupied nuclear states and pick-up to investigate occupied states.

For example let us consider the reaction ${}_6C^{12}$ (d, p) ${}_6C^{13}$. According to the shell model C^{12} has spin 0 and positive parity and C^{13} has spin 1/2 and negative parity. By the use of parity conservation law

$$(-1) = +(1)(-1)^{l_c}.$$

Thus l_c of the captured particle must be odd.

We know that the angular momentum of the captured particle

$$I_c = I_f - I_i = \frac{1}{2} - 0 = \frac{1}{2}.$$

Since $\quad I_c = I_c + s_c \quad$ or $\quad I_c = I_c - s_c$

$$\therefore \quad I_c = \vec{\frac{1}{2}} - \vec{\frac{1}{2}} = 0 \text{ or } 1.$$

The $l_c = 0$ is not acceptable as it violates parity law, hence $l_c = 1$. Thus if the shell model prediction of C^{13} spin and parity is correct, then we should observe a peak around 17° in the differential scattering cross-section corresponding to $l = 1$ capture. This is exactly what is observed experimentally.

In the reaction P^{31} (d, p) P^{32}, the target nucleus has angular momentum $I_i = \frac{1}{2}^+$ (an odd $2s_{1/2}$ proton) and the ground state of P^{32} is expected to be formed by adding a neutron in a $d_{3/2}$ state. The two l-values can combine to $I_f = I_p + I_n = 1^+$ or 2^+. If we apply the model independent selection rule we find that the ground state (1^+ state) can be reached both by an $l = 0$ and $l = 2$ neutron, whereas the shell model state $d_{3/2}$ requires that $l = 2$. Bethe and Butler raised the question : Will a stripping experiment yield an angular distribution consistent with $l = 2$, with $l = 0$ or with a mixture of these?

In the same year, after performing experiment, Parkinson, Beach and King found an angular distribution consistent with $l = 2$ and very little (if any) trace of $l = 0$ contribution. In this particular example the test is very sensitive because $l = 0$ stripping gives a cross-section that is one order of magnitude larger at the maximum than does $l = 2$ stripping under similar circumstances. The result is, therefore, a great tribute to the shell model.

INVERSE PROCESS

(Reciprocity Theorem): Let us consider a reversible process X + α = Y + y, in which X, x, Y and y occur in arbitrary numbers in a large box of volume V. We are interesting in the relation between the total cross-section $\sigma(x \rightarrow y)$, most generally $\sigma(\alpha \rightarrow \beta)$ of the reaction with entrance channel α and reaction channel β and the total cross-section $\sigma(\beta \rightarrow \alpha)$ of the inverse reaction. For this we use the fundamental theorem of statistical mechanics *(the principle of overall balance),* which states that when the system is in dynamical equilibrium all energetically permissible states are occupied with equal probability. Here we are interested in two particular states, the reaction channels α and β. The theorem is then equivalent to stating that in a given energy range the number of possible channels in the box is proportional to the number of possible channels into the box. The latter is given by

$$N_\alpha = \frac{4\pi p_\alpha^2 \ V \ dp_\alpha}{h^3} = \frac{p_\alpha^2 \ V \ dp_\alpha}{2\pi \hbar^3}.$$

Since $\quad v = \frac{dE}{dp}$, hence $N_\alpha = \frac{p_\alpha^2 \ VdE_\alpha}{2\pi^2 \hbar^3 v_\alpha}$. ...(1)

Similarly, we have $N_\beta = \frac{p_\beta^2 \ VdE_\beta}{2\pi^2 \hbar^3 v_\beta}$. ...(2)

The energy range for the two channels must of course be the same, *i.e.,* $dE_\alpha = dE_\beta$, hence

$$\frac{\text{No. of channels } \alpha \text{ in the box}}{\text{No. of channels } \beta \text{ in the box}} = \frac{N_\alpha}{N_\beta} = \frac{p_\alpha^2 \ v_\beta}{N_\beta \ v_\alpha}. \qquad ...(3)$$

The system is in dynamical equilibrium when the number of the transitions $\alpha \rightarrow \beta$ per second is equal to the number of transitions $\beta \rightarrow \alpha$ per second. The condition usually holds and is known as the *principle of detailed balance.* Further

No. of transitions $\alpha \rightarrow \beta$ per sec = $N_\alpha \times \omega\ (\alpha \rightarrow \beta)$,

where $\omega(\alpha \rightarrow \beta)$ is the transition probability for the reaction $(\alpha \rightarrow \beta)$.

Hence $\quad p\alpha^2 \ v_\beta \ \omega(\alpha \rightarrow \beta) = p\beta^2 \ v_\alpha \ \omega(\beta \rightarrow \alpha)$. ...(4)

The transition probability measures the chance that one particle moving with velocity v in volume V is scattered per sec. Hence the cross-section σ which corresponds to unit incident flux is given by the relation

$$\sigma = \frac{\omega V}{v}. \qquad ...(5)$$

Combining relations (4) and (5) and using $k = \frac{p}{\hbar}$, we have

$$k\alpha^2\, \sigma(\alpha \to \beta)\, k\beta^2\, \sigma(\beta \to \alpha) \qquad ...(6)$$

or
$$\frac{\sigma(\alpha \to \beta)}{\lambda\!\!\!^{-}\alpha^2} = \frac{\sigma(\beta \to \alpha)}{\lambda\!\!\!^{-}\beta^2}. \qquad ...(7)$$

We have assumed zero intrinsic angular momenta for the particles so far. If I is the intrinsic angular momentum of any one of the particles, the corresponding density of states then must be multiplied by 2I + 1. This if there are intrinsic momenta for X, x, Y and y, we may write

$$(2I_X + 1)\,(2I_x + 1)\, k_\alpha^2 \sigma(\alpha \to \beta) = (2I_Y + 1)\,(2I_y + 1)\, k_\beta^2 \sigma(\beta \to \alpha).$$

...(8)

If the initial and final states have definite angular momenta then the above equation must be employed.

CROSS-SECTION OF A NUCLEAR REACTION

Bohr assumed that the mode of disintegration of a compound system depends only on its energy, angular momentum and parity, but not on the specific way in which it has been produced. Therefore, to calculate the cross-section of a nuclear reaction, it is necessary to determine the cross-section of the two processes, the formation of the compound nucleus and its decay.

$$X + x \to C \to Y + y. \qquad ...(1)$$

Since the break up process of the compound nucleus is independent of the formation process, hence we can write in general

$$\sigma(\alpha, \beta) = \sigma_C(\alpha)\, G_C(\beta), \qquad ...(2)$$

where σ(α, β) is the cross-section corresponding to a specific entrance channel α and specific exit channel β, $\sigma_C(\alpha)$ is the cross-section for the formation of a compound nucleus C with particle x through channel α and $G_C(\beta)$ is the probability that the compound nucleus C will disintegrate by emitting a particle y through channel β.

For the sake of simplicity we shall neglect the dependence of $G_C(\beta)$ on angular momentum and parity. $G_C(\beta)$ then only depends on the excitation energy (E_C) of compound nucleus. Because of the Heisenberg uncertainty principle this energy cannot be sharply defined. The uncertainty in excitation is measured by a quantity Γ_β, the level width associated with the decay constant λ_β of compound system.

$$\Gamma_\beta \text{ (For } E_C) = \hbar\lambda_\beta = \hbar\tau_\beta \text{ (for } E_C), \quad ...(3)$$

where τ_β is the decay time of the compound nucleus through the channel β. If we add all the decay constants for the possible exist channels then we get the total decay constant for the compound nucleus. Therefore, the total width of the compound level with an excitation energy E_C is given by

$$\Gamma(\text{for } E_C) = \sum_\beta \Gamma_\beta(\text{for } E_C). \quad ...(4)$$

The *branching probability,* the relative probability for the compound nucleus at excitation energy E_C to break up by emitting particle y, in terms of decay rates is given by

$$G_C(\beta) = T_\beta \text{ (for } E_C)/T(\text{For } E_C). \quad ...(5)$$

The assumption of the independence of two processes (the creation and the disintegration of the compound system) gives a relation, which depends only in the kα and kβ. It is derived earlier and can be rewritten as

$$k\alpha^2\, \sigma(\alpha, \beta) = k\beta^2\, \sigma(\beta, \alpha). \quad ...(6)$$

Using eqns (2) and (5), this relation becomes

$$k\alpha^2 \frac{\sigma_C(\alpha)\,\Gamma_\beta\,(\text{for } E_C)}{\Gamma(\text{for } E_C)} = k\beta^2 \frac{\sigma_C(\beta)\,\Gamma_\alpha(\text{for } E_C)}{\Gamma(\text{for } E_C)}$$

or

$$\frac{k\alpha^2\,\sigma_C(\alpha)}{\Gamma_\alpha\,(\text{for } E_C)} = \frac{k\beta^2\sigma_C(\beta)}{\Gamma_\beta\,(\text{for } E_C)} = U(E_C). \quad ...(7)$$

This proportional coefficient $U(E_C)$ is a function depending upon the excitation energy E_C of the compound nucleus only, but not on the particular channel used. This relation shows that there is a simple relationship between the probability for breakup and the cross-section for forming the compound nucleus through the same channel. The probability of decay through a channel β

$$G_C(\beta) = \frac{\Gamma_\beta}{\Gamma} = \frac{k\beta^2 \sigma_C(\beta)}{\left[\Gamma_\alpha + \Gamma_\beta + ...\right] U(E_C)} = \frac{k\beta^2 \sigma_C(\beta)}{\sum_\gamma k_\gamma^2 \sigma_C(\gamma)}, \quad ...(8)$$

where the sum is extended over all channels γ into which compound system can decay. It is, therefore, possible to derive the decay probability through a given channel of the compound nucleus if the cross-section $\sigma_C(\gamma)$ for the formation of the compound nucleus by all possible channels are known, and if the Bohr assumption is valid. The validity of Bohr's hypothesis of the compound nucleus has been elegantly demonstrated by several experiments. Let us discuss the experiment of Indian Scientist S.N. *Ghoshal* (1950) in which he produced the same compound nucleus Zn^{64*} with the bombardment of $_{28}Ni^{60}$ by α-particle and $_{29}Cu^{52}$ by protons. Additional 7 MeV was added to the K.E. of protons to produce the same excitation in Zn^*. The reactions observed were

Ni^{60} (α, n) Zn^{63}	Cu^{63} (p, n) Zn^{63}
Ni^{60} (α, 2n) Zn^{62}	Cu^{63} (p, 2n) Zn^{62}
Ni^{60} (α, pn) Cu^{62}	Cu^{63} (p, pn) Cu^{62}.

Since the excitation produced through these two processes are the same, the disintegration probability $G_C(\beta)$ is the same, because it depends upon the excitation produced in the compound nucleus and not upon the mode of formation. If the compound nucleus theory is true, then one should expect.

$$\sigma(p, n) : \sigma(p, 2n) : \sigma(p, pn) = \sigma(\alpha, n) : \sigma(\alpha, 2n) : \sigma(\alpha, pn). \quad ...(9)$$

Fig. 2.9 shows measurements due to Ghoshal, here six cross-sections are plotted as functions of K.E. of the β-particle and protons. The proton-energy scale has been shifted by 7 MeV with respect to the α-particle energy scale in order to make the peaks of the proton curves correspond with those of the α-particle curve. It can be seen from figure that the results are accurate with in 10%, and thus providing evidence for the validity of the compound nucleus theory.

The small difference between experimental and theoretical values can be explained as under : Ghoshal's experiment was performed at high bombarding energies. The levels excited in the compound nucleus are more closely spaced and broader and may overlap partially. This can be explained by the *continuum theory.* Thus we see that the compound nucleus theory given here is applicable to low bombarding energy.

NUCLEAR REACTIONS

When a target nucleus is bombarded by means of energetic projectile particles. The bombardment process may give rise to the following events single or jointly.

(a) Elastic Scattering

The incident particle strikes the target nucleus and leaves without energy loss but, in general, with altered direction of motion. Scattering of α-particles in gold is a good example of this process.

(i) $e^- + {}_6C^{12} \rightarrow {}_6C^{12} + e^-$

(ii) ${}_2He^4 + {}_{79}Au^{197} \rightarrow {}_{79}Au^{197} + {}_2He^4$

(iii) ${}_0n^1 + {}_{92}U^{238} \rightarrow {}_{92}U^{238} + {}_0n^1$

(b) Inelastic Scattering

The scattered particle may loss K E in excess of that required for an elastic collision with the nucleus. This being corresponding increase in the internal energy of the nucleus which is excited to a higher quantum state. This inelastic scattering can be represented by the well known example.

(i) ${}_0n^1 + {}_{82}Pb^{208} \rightarrow {}_{82}Pb^{208*} + {}_0n^1$

(ii) ${}_3Li^7 + {}_1H^1 \rightarrow ({}_3Li^7)^* + {}_1H^1$

(iii) ${}_1H^1 + {}_7N^{14} + 7N^{14*} + {}_1H^1.7N^{14*} \rightarrow 7N^{14} + y.$

The star (*) is used to indicate that, after scattering, nucleus is left in an excited state. In this present example the excess energy is latter radiated away in the from of a γ-quantum.

(c) Radiative Capture

A particle may combine with a nucleus to produce a new nucleus or a compound nucleus which is in an excited state. The excess energy is emitted in the form of γ-ray photons. This type of process is known as radiative capture example

$${}_{12}Mg^{26} + {}_1H^1 \rightarrow ({}_{13}Al^{27})^* \rightarrow {}_{13}Al^{27} + \gamma.$$

(d) Disintegration

On striking the target nucleus the incident particle is absorbed and a different particle is ejected. The product nucleus differs from target nucleus. The incident particle may be an α-particle, proton, neutron, etc. The product particle may be a charged particle or a neutron. Thus process is known disintegration as

$$_7N^{14} + _2He^4 \rightarrow _3O^{17} + _1H^1.$$

(e) Photo Disintegration

The γ-rays are absorbed by the target nucleus, exciting it to a higher quantum state. If the energy is high enough, one or more particles may be liberated. An example is

$$_1H^2 + g \rightarrow _1H^1 + _0n^1.$$

(f) High Energy Reactions

In the energy range about 150 MeV, spallation process merges into new kinds of reactions in which new kinds of particles (mesons, strange particles) are produced along with neutrons and protons.

(g) Direct Reactions

A collision of an incident particle with the nucleus may immediately pull one of the nucleons out of the target nucleus by the so called *pick-up reaction.* In the inverse process, a bombarding particle composed of more than one nucleon may loss one of them to the target by the so-called *stripping reaction, e.g.*

$$_{29}Cu^{63} + _1d^2 \rightarrow _{29}Cu^{64} + _1H^1.$$

(h) Heavy Ion Reactions

Nuclear reactions induced by heavy ions exhibit the characteristics both of compound nucleus and of the stripping and pick up mechanisms. Example

$$N^{14} + Pb^{207} \rightarrow N^{13} + Pb^{203}.$$

(i) Spontaneous Decay

β– and α – decay processes may be regarded as this type of nuclear reactions. In these reactions the total energy of the system is not under the experimenter's control.

(j) Spallation Reactions

On capture a incident particle, a heavy nucleus has sufficient energy for the ejection of several particles. Such a reaction is known as spallation reaction, *e.g.*, the nuclear fission, in which heavy nucleus splits mainly in two nuclei,

Example : ${}_{92}U^{235} + {}_{0}n^{1} \rightarrow {}_{40}Zr^{98} + {}_{52}Te^{136} + 2{}_{0}n^{1}$.

CONSERVATION LAWS

We shall merely list the various conservation laws that appear to be valid in ordinary nuclear interactions.

1. Conservation of Angular Momentum

The total angular momentum *l* comprising the vector sum of the intrinsic angular momentum s and relative orbital momentum *l* of the products must be equal to the total angular momentum of the initial particles.

2. Conservation of Energy

The total energy of the products, including both mass energy and kinetic energy of the particles plus the energy involved must be equal to the mass energy of the initial ingredients plus the kinetic energy of the bombarding particle.

3. Conservation of Momentum

The total linear momentum of the products must be equal to the linear momentum of the bombarding particle (the target nucleus is ordinarily taken to be at rest).

4. Conservation of Isotopic Spin

The invariance of the nuclear Hamiltonian function towards the charge character of the nucleons can be expressed analytically as an invariance towards rotational shifts of the axes in isotopic spin space, and there should correspondingly exist a conservation law for the isotopic spin of a nuclear system.

5. Conservation of Parity

The parity of the system determined by the target nucleons and bombarding particle must be conserved throughout the reaction. The total parity of the system is the product of intrinsic-parities of the target nucleus and bombarding nucleus. No violation of parity has been observed in a nuclear reaction (strong nuclear forces). Although parity does not appear to be conserved in weak interactions.

6. Conservation of Spin

The important conservation law asserts that the spin character or a closed system cannot change, *i.e.*, the statistics remains same as that existed before reaction.

7. Conservation of Nucleons

The law of conservation of nucleons states that the nucleons can neither be created nor can be destroyed so that the number of nucleons minus the number of anti-nucleons in the universe remains constant.

8. Conservation of Charge

The total electric charge of the products must be equal to the total electric charge of initial particles.

CLASSICAL ANALYSIS OF CROSS-SECTION

If the bombarding particle has an impact parameter (the distance of closest approach) b with respect to the nucleus. The angular momentum of the system will be normal to the relative momentum p of the two particles and of magnitude

$$L = pb = \frac{\hbar b}{\lambda\!\!\!^{-}}.$$

If b > R, the bombarding particle should not have much effect on the target nucleus as it is outside the range of nuclear forces. Thus we expect the most important nuclear interactions to occur with those bombarding particles which have angular momentum less than or equal to the maximum value $\frac{\hbar R}{\lambda\!\!\!^{-}}$. Quantum mechanics requires that the angular momentum is quantized and can adopt the following discrete values $\hbar l$ with $l = 0, 1, 2...$ If we now consider the associated wave $\lambda\!\!\!^{-}$ we see that it may be resolved into partial waves, each corresponding to a value of l. Blatt and Weisskopf have provided the significance of the angular momenta in a plane wave by dividing up the incident beam into cylindrical zones, such that each zone is associated with a different impact parameter. In the central zone there will only be the particles with an impact parameter b less than $\lambda\!\!\!^{-} = \left(= \frac{\hbar}{p} \right)$. Actually it is not possible to visualize an impact parameter less than the associated wavelength (*Uncertainly principle*). The angular momentum is zero throughout this zone. The next zone contains all particles with impact parameters between $\lambda\!\!\!^{-}$ and $2\lambda\!\!\!^{-}$. Similarly the lth zone contains particles with impact parameters between $l\lambda\!\!\!^{-}$ and $(l + 1)\lambda\!\!\!^{-}$. The change in angular momentum (*i.e.*, momentum × impact parameter) on passing from zone l to $l + 1$ is $\hbar l$ to $h(l + 1)$. The area of the lth zone is obtained from the difference between the areas due to radii $(l + 1)\lambda\!\!\!^{-}$ and $l\lambda\!\!\!^{-}$. This is the cross-sectional area σ_r^l of the lth zone. If it is assumed that each particle hitting the nucleus causes a reaction then the reaction cross-section may be obtained by summing σ_r^l, over all values of l from 0 to l_m

$$\therefore \qquad \sigma_r = \sum_0^{l_m} \pi\lambda\!\!\!^{-2}(2l+1) = \pi\lambda\!\!\!^{-2}(l_m+1)^2. \qquad ...(1)$$

The maximum value of l, corresponds to maximum impact parameter R, is given by $l_m = R/\lambda\!\!\!^{-}$. Substituting this value of l_m in the above relation we have maximum reaction cross-section

$$\sigma_r = \pi (R + \lambda\!\!\!^{-})^2. \qquad ...(2)$$

This result suggests the possibility of nuclear reaction cross-sections which are several orders of magnitude larger than the geometrical cross-section of the nucleus.

However, it will be recalled that the incident wave may also be partially reflected and that a transmission factor Tl, depending on the value of l, must be taken into account. The quantum mechanical treatment of the problem gives

$$\sigma_r = \pi\lambda\!\!\!^{-2} \sum_{l=0}^{\infty} (2l+1)T_l. \qquad ...(3)$$

An interesting case is that of a nucleus which has a radius R very much greater than $\lambda\!\!\!^{-}$, *i.e.*, when the high energy particles strike the nucleus. The nucleus is considered as completely absorbent (black) which corresponds to $T_l = 1$. It is for all values of l upto and including $l_m (= R/\lambda\!\!\!^{-})$. In these conditions the cross-section obtained is same as shown by eqn. (2). This is close to the geometrical cross-section when $\lambda\!\!\!^{-}$ is small is comparison to R. At very high energies, the nucleus exhibits a certain transparency to particles which may penetrate it without interaction. For the weaker energies, $\lambda\!\!\!^{-}$ is too large compared with R for the simplifying hypothesis $T_l = 0$ for large l to hold, since even for $l = 0$, $l > R$.

The elastic cross-section depends upon the condition under which the incident wave is reflected. The interference with the incident wave gives rise to amplitudes four times that of the penetrating wave.

$$\sigma_{el}(\max) = 4\pi(R + \lambda\!\!\!^{-})^2. \qquad ...(4)$$

In contrast, when $T_l = 1$ for all the values of l, in the case of the black nucleus previously examined, the minimum value is given as

$$\sigma_{el}(\min) = \pi(R + \lambda\!\!\!^{-})^2 = \sigma_{inel}. \qquad ...(5)$$

The effective total cross-section is, therefore, equal to

$$\sigma_l = \sigma_{el} + \sigma_{inel} = 2\pi\,(R + \lambda\!\!\!^{-})^2. \qquad ...(6)$$

Thus we see that for black nucleus the total cross-section is twice the geometrical cross-section. The opaque disc of the nucleus projects a shadow and diffraction occurs as in wave optics. If the nucleus is replaced by a disc of radius $(R + \lambda\!\!\!^{-})$ this shadow may be shown to exist. The disc then emits particle waves of the same intensity in the direction of the incident beam but opposite in phase to those of the beam.

In the derivation of above theory we have not considered the effect of Coulombs repulsion. The Coulomb repulsion will bring the relative K.E. of the charged particle from T (when particle is far apart from nucleus) to T – B, when the particle is just touching the nucleus of radius R. If z and Z are the atomic numbers of incident particle and target nucleus respectively, then $B = zZe^2/4\pi\epsilon_0 R$. Thus the Coulomb repulsion slows down the particle so that the initial moment $(2MT)^{1/2}$ is decreased to $[2M(T - B)]^{1/2}$, where M is the reduced mass of the system. Hence the maximum angular momentum.

$$L_m = R[2M(T - B)]^{1/2} = R(2MT)^{1/2}(1 - B/T)^{1/2}$$

$$\therefore \qquad b_m = \frac{L_m}{p} = \frac{L_m}{(2MT)^{1/2}} = R\left(1 - \frac{B}{T}\right)^{1/2}$$

Thus the Coulomb repulsion diminishes the l_m by a factor of $\left(1 - \frac{B}{T}\right)^{1/2}$. The upper limit for the capture of charged particles can be estimated as the area of the disc of radius b_m.

$$\sigma_r = \pi R^2\left(1 - \frac{B}{T}\right). \quad ...(7)$$

PARTIAL-WAVE ANALYSIS OF REACTION CROSS-SECTION

Consider an incident plane wave of neutrons travelling parallel to the z-axis. This plane wave has a spatial part e^{ikz}, where $k = 2\pi/\lambda$, and can be expressed as a super-position of spherical waves with different orbital angular momentum quantum numbers relative to the target nucleus. The expansion is

$$e^{ikz} = e^{ikr\cos\theta} = \sum_{i=0}^{\infty} i^i \sqrt{[4\pi(2l+1)]}\, j_l(kr)\, Y_{l,0}(\theta) \quad ...(1)$$

This solution involves Spherical Bessel functions j_l (kr) and spherical harmonics $Y_{l,m}(\theta)$. As there is cylindrical symmetry about the z-axis the order m of the spherical harmonics is zero. If r, the distance of the incident particle from the target nucleus, is very large so that $kr \geq l$, then the form of the function j_l (kr) is

$$j_i(kr) = \frac{\sin\left(kr - \frac{1}{2}l\pi\right)}{kr} = i\,\frac{e^{-i(kr - l\pi/2)} - e^{+(kr - l\pi/2)}}{2kr}.$$

$$\therefore \quad e^{ikz} = \frac{\sqrt{\pi}}{kr}\sum_{l=0} \sqrt{(2l+1)i^{l+1}\left[e^{-i(k_r - l\pi)} - e^{+i(k_r - l\pi/2)}\right]}\, Y_{l,0}(\theta) \quad ...(2)$$

The first exponential term represents a series of spherical waves coming towards the origin with momentum $k\hbar$ and angular momentum $l\hbar$, the second exponential term represents a series of a outgoing spherical waves with just the same properties. The above expression describes an undisturbed plane wave.

If however, a nucleus is located at the origin, the amplitudes and phases of the outgoing spherical waves from the origin are, in general, changed. Let the outgoing part of the partial wave with orbital angular momentum l be changed by a factor η_l. This may be a complex number. This is related to the phase shift by $\eta_l = |\eta_l|\, e^{2i\sigma l}$. Elastic scattering will take place, when $|\eta_l| = 1$. If $|\eta_l| < 1$ both elastic scattering and nuclear reaction will take place. Thus the wave function for the disturbed wave is

$$\phi(r) = \frac{\sqrt{\pi}}{kr}\sum_{l=0}^{\infty}\sqrt{(2l+1)i^{l+1}\left[e^{-i(k_{r-l\pi/2})-\eta_l}e^{+i(k_{r-l\pi/2})}\right]}Y_{l,0}(\theta) \quad ...(3)$$

The elastically scattered wave function ψ_{sc}, which is the difference between the disturbed wave and the incident wave, is given by

$$\psi_{sc} = \psi(r) - e^{ikx}$$

$$= \frac{\sqrt{\pi}}{kr}\sum_{l=0}^{\infty}\sqrt{(2l+1)i^{l+1}(1-\eta_l)e^{i(kr-l\pi/2)}}\,Y_{l,0}(\theta) \quad ...(4)$$

Since $\psi(r) \propto e^{ikz} + f(\theta)\, e^{ikr}.r$, hence

$$f(\theta) = \frac{\sqrt{\pi}}{k}\sum_{l=0}^{\infty}\sqrt{(2l+1)i^{l+1}(1-h_l)e^{-il\pi/2}}\,Y_{l,0}(\theta).$$

The scattered cross-section is obtained by dividing the number N_{sc}. of scattered particles per second by the number of incident particles per unit area per second which for the plan wave e^{ikx} is v, the velocity if the particles. Consider a sphere of radius r_0. If r_0 is very much larger than the range of the nuclear forces, than the number of scattered particles per second N_{sc} is equal to the flux of ψ_{sc} this sphere.

$$\therefore\ N_{sc} = \frac{\hbar}{2iM}\int\left(\frac{\partial\psi_{sc}}{\partial r}\psi^*_{sc} - \frac{\partial\psi^*_{sc}}{\partial r}\psi_{sc}\right)_{r=r_0} r_0^2 \sin\theta\, d\theta\, d\phi, \quad ...(5)$$

where M is the mass of the incident particle. Here integral is extended over the surface of the sphere. Substituting values of ψ_{sc} and ψ^*_{sc},

$$N_{sc} = \frac{\hbar}{2iM}\int\left[f^*(\theta)\frac{e^{-ikr0}}{r_0}f(\theta)\left(ik\frac{ikr_0}{r_0}\frac{1}{r_0}\frac{e^{ikr_0}}{r_0}\right)\right.$$

$$\left.-(\theta)\frac{e^{ikr^0}}{r_0}f^*(\theta)\left(-ik\frac{e^{ikr_0}}{r_0}\frac{1}{r_0}\frac{e^{ikr_0}}{r_0}\right)\right]r_0^2 \sin\theta\, d\theta\, d\phi$$

$$= \frac{h2ik}{2iM}\int f^*(\theta)\, f(\theta).\frac{1}{r_0^2}r_0^2 \sin\theta\, d\theta\, d\phi$$

$$\because\ \frac{\hbar k}{M} = \left(\frac{h}{2\pi}\right)\left(\frac{2\pi}{\lambda}\right)\left(\frac{1}{M}\right) = \frac{p}{M} = v,$$

$$\therefore \quad N_{sc} = \frac{\hbar k}{M}\int |f(\theta)|^2 \sin\theta \, d\theta \, d\phi = v\int |f(\theta)|^2 d\Omega \qquad ...(6)$$

Because of the orthogonality and normalization of the $Y_{l,0,}$ we get

$$N_{sc} = v\frac{\pi}{k^2}\sum_{i=0}^{\infty}(2l+1)|1-\eta_l|^2 \qquad ...(7)$$

Thus the scattering cross-section is given by

$$\sigma_{sc} = \frac{N_{sc}}{v} = \frac{\pi}{k^2}\sum_{i=0}^{\infty}(2l+1)|1-\eta_l|^2 \qquad ...(8)$$

If we define

$$\sigma_{l,\,sc} = \pi\lambda\!\!\!^{-2}\,(2l+1)\,|1-\eta_l)^2, \qquad ...(9)$$

then eqn. (8) can be rewritten as

$$\sigma_{sc} = \sum_{i=0}^{\infty} s_{l,\,sc}\,. \qquad ...(10)$$

Thus we see that in the integral cross-sections the contributions of the partial waves for different l are simply added.

We now derive an expression for the reaction cross-section σ_{Re}. It is determined by the number N_{Re}, the number of particles which enter a sphere of radius r_0 around the enter without leaving it again through the entrance channel. It is, therefore, equal to the net flux into this sphere as computed from the complete wave-function $\psi(r)$.

$$N_{Re} = -\frac{h}{2iM}\int\left(\psi^*.\frac{\partial\psi}{\partial r}\psi\frac{\partial\psi^*}{\partial r}\right)_{r=r0} r_0^2 \sin\theta\, d\theta\, d\phi \qquad ...(11)$$

The minus sign is used. because inward flux N_{Re}, is positive.

Using the orthogonality and normalization of the Y_l, 0, we get

$$N_{Re} = v\frac{\pi}{k^2}\sum_{i=0}^{\infty}(2l+1)\left(1-|\eta_l|^2\right) \qquad ...(12)$$

$$\therefore \quad \sigma_{Re} = \frac{\pi}{k^2}\sum_{i=0}^{\infty}(2l+1)\left(1-|\eta_l|^2\right) = \pi\lambda\!\!\!^{-2}\sum_{i=0}^{\infty}(2l+1)\left(1-|\eta_l|^2\right). \qquad ...(13)$$

Using the definition

$$\sigma_{l,\,Re} = \pi\lambda\!\!\!^{-2}\,(2l+1)\,(1-|\eta_l|^2). \qquad ...(14)$$

We can represent eqn. (13) as $\sigma_{Re} = \sum_{i=0}^{\infty} \sigma_{l, Re}$. ...(15)

The total cross-section for a given l is the sum of equations (9) and (14)

$$\sigma_{l, t} = \sigma_{l, sc} + \sigma_{l, Re}$$
$$= \pi \lambda\!\!\!^{-2} (2l+1)\left[\left|1-\eta_l\right|^2 \_ \left(1-\left|\eta_l\right|^2\right)\right]$$
$$= 2\pi \lambda\!\!\!^{-2} (2l + 1) [1 - R_e(\eta_l)], \quad ...(16)$$

where $R_e(\eta_l)$ is the real part of the complex number η_l.

Conclusion

1. If $\eta_l = 0$, then $s_{l, Re}$ takes on its maximum value $\pi \lambda\!\!\!^{-2}(2l+1)$, however the scattering cross-section takes on the finite value. Scattering and reaction cross-sections are then both of the same size.
2. If $\eta_l = -1$, then $\sigma_{l, Re} = 0$ and the scattering cross-section $\sigma_{l, sc}$ is maximum having value $4\pi \lambda\!\!\!^{-2}(2l+1)$.
3. If $h_1 = 1$, then $\sigma_{l, sc} = 0$. It also implies that there are no reactions, as there is no case in which reactions occur without elastic scattering.
4. If $|\eta_l| = 1$, then $\sigma_{l, Re} = 0$, but $\sigma_{l, sc}$ may be finite for the complex value of η_l.
5. $|\eta_l|^2$ can never be negative, otherwise the outgoing wave will have a high intensity than the incoming one. Thus the condition $|\eta_l|^2 \leq 1$ insures that the cross-section $\sigma_{l, Re}$ does not become negative.

If the incident neutron has a high enough energy, the reduced wavelength $\lambda\!\!\!^{-}$ is small compared to the radius R of the nucleus. Suppose the nucleus is completely black to the incident neutrons, *i.e.*, all neutrons which hit the nucleus are absorbed by it and, therefore, lead to a reaction, all the particles with $l \leq R/\lambda\!\!\!^{-}$ strike the nucleus. We can express this as

$$\eta_l = 0 \text{ if } l \leq \frac{R}{\lambda\!\!\!^{-}} \text{ and } \eta_l = 1 \text{ if } l > \frac{R}{\lambda\!\!\!^{-}}. \quad ...(17)$$

Therefore all incident particles with $l \le \frac{R}{\lambda}$ react and all those with $l > \frac{R}{\lambda}$ pass by. Thus eqn. (13) gives

$$\sigma_{Re} = \sum_{i=0}^{R/\lambda} \pi\lambda^2 (2l+1) = \pi (R+\lambda)^2. \quad ...(18)$$

When σ_{Re} has its maximum value, $\sigma_{Re} = \sigma_{Re}$, and thus we obtain the apparently paradoxical result that the total cross-section is twice them geometrical cross-section of the nucleus:

$$\sigma_l = \sigma_{Sc} + \sigma_{Re} = 2\pi(R+\lambda)^2 \sim 2\pi R^2. \quad ...(19)$$

The physical meaning of the results involves the diffraction of the incident wave to form a shadow behind the absorbing centre. Since any neutron striking the nucleus reacts, therefore, the nucleus will appear opaque to a fast neutron.

COMPOUND NUCLEUS

According to the Bohr, the nuclear reaction takes place in two distinct and independent stages :

(i) *formation of a compound nucleus C and*

(ii) *the disintegration of the compound nucleus into the products of the reaction.* The compound nucleus, which is a many body system of strongly interacting particles, is formed by the amalgamation of an incident particle x with a target nucleus X.

$$X + x \to C^*. \quad ...(1)$$

The incident particle captured by a nucleus gives up its energy to few nucleons and, as the results of the interaction of these nucleons with all the others, the energy is quickly distributed among all the nucleons of the compound nucleus. The new nucleus thus formed is in excited state. If E_x is the kinetic energy of the incident particle of mass m, the excitation energy of the compound nucleus will be given by

$$E^* = \frac{E_x M}{(m+M)+E_B}, \quad ...(2)$$

where M is the mass of the target nucleus and E_0 the binding energy of the particle in the compound nucleus. The mode of disintegration of compound nucleus ($C^* \rightarrow Y + y$) is independent of the mode of formation and depends only on its energy, angular momentum and parity. If the different processes lead to the same compound nucleus the decomposition is identical. A compound nucleus, once formed, can decay in a number of different ways, each with its own-intrinsic probability. For example a compound nucleus Al^{27*}, formed by the bombardment of Na^{23} with α-particles or by other methods, can decay at least in following ways : $Al^{27*} \rightarrow (Na^{23} + \alpha), (Mg^{25} + d), (Mg^{26} + p), (Al^{26} + n)$ or $(Al^{27} + \lambda)$. In each case the residual nucleus can usually be left in one of its excited states, leading to still more modes of decay.

The compound nucleus, has a life time which is long ($10^{-14} - 10^{-18}$ sec) compared to the time for a nucleon to traverse a nucleus ($10^{-20} - 10^{-23}$ sec.). This finite life time is because there can always be a statistical fluctuation in the energy distribution which concentrates enough energy on a nucleon to allow to escape.

Energy Levels of Nuclei

Resonances in the yield curve of a nuclear transmutation will occur whenever the energy of the incident particle together with its B.E. in the compound nucleus is equal to and coincides with an energy level of that nucleus. An energy level that corresponds to an excitation energy greater than the B.E. of the particle in the compound nucleus is a *virtual energy level* and the nucleus is than said to be in a *quasi-stationary state.* The most stable energy state of nucleus is the lowest energy level or *ground state* (*zero potential energy*). Corresponds to this state the energy of the nucleus is the B.E. calculated from the mass defect. For excitation energies less than the B.E., the corresponding energy levels are *bound energy levels*. In the normal state of the nucleus, the lowest bound levels are all occupied. The nucleon will be in one of the normally unoccupied bound states, if the excitation energy is less than about 8 MeV. The region of the virtual levels, contains those energy levels that can be occupied during a nuclear reaction while the nucleus exists in an intermediate state as a compound nucleus. De-excitation of bound energy levels can take place only by emission of γ-rays, whereas de-excitation from a virtual energy level can occur in a variety of ways, either by particle emission or γ-emission. For example, in certain (α, p) reactions discrete energy values are observed amongst the ejected protons, although monoenergetic α-particles are being used in the bombardment of the target. These experimental results can be understood if one attributes to èach energy group a corresponding energy level in the nuclei of their origin.

Thus, the largest of the group energies E_0 is associated with the formation of residual nucleus in its lowest or ground state. The smaller energy values (E_1, E_2, E_3) of the groups are associated with residual nuclei which have been left in higher excited energy states. The energy differences between the various groups can be related to the respective energy level spacing in the residual nucleus. This conclusion is supported by the fact that the lower energy groups are accompanied by simultaneous γ-ray emission, whereas the highest energy group is not accompanied by any γ-emission from the residual nucleus, as shown in Fig. 2.8.

Level Width and De-excitation

The excited state of the compound nucleus has a definite mean life time before it decays by one of the possible modes of de-excitation. We know that

Mean life time τ = 1/disintegration constant λ, ...(3)

where λ is connected with the level width Γ by the relation

$$\Gamma = \hbar\lambda \qquad ...(4)$$

$$\therefore \qquad \Gamma = \frac{\hbar}{\tau} \text{ or } \Gamma\,\tau = \hbar. \qquad ...(5)$$

According to the principle of Heisenberg uncertainty

$$\Delta E.\Delta t = \hbar. \qquad ...(6)$$

If we equate the uncertainty in the time measurement Δ t with the mean life time of the excited state, we obtain level width Γ for the uncertainty in the energy of the excited state Δ E. *Hence the level width of the excited state is a spread in energy of the excited level due to the uncertainty in the level.* As the level width is inversely proportional to the mean life time τ, it is clear that a long life time means a very fine and narrow energy level, whereas a short life time means a broad or diffuse energy level of larger level width.

For each individual mode of decay, there is a different probability of decay and, therefore, a different partial width for each decay product. The total width of an energy level is then the sum of the individual partial widths, *i.e.*, $\Gamma = \Gamma_1 + \Gamma_2 + \Gamma_3 + ...$

For bound levels, neglecting internal conversion $\Gamma = \Gamma_\gamma$. But for virtual level $\Gamma = \Gamma_\gamma + \Sigma\Gamma_{particles}$, where the summation includes the widths corresponding to the emission of certain energetically allowed particles. The widths Γ_γ and $\Gamma_{particles}$ are partial widths. For the ground state $\Gamma = 0$.

NUCLEAR REACTION KINEMATICS

There are certain restrictions in the conservation of energy and momentum. These restrictions are called kinematic restrictions and this mathematical method is known as *kinematics*. Consider a nuclear reaction

$$X + x \rightarrow Y + y, \qquad ...(1)$$

where X, x, Y and y are the target nucleus, bombarding particle, product nucleus and product particle, respectively. It will be assumed that target nucleus is at rest so it has no kinetic energy. Since total energy is conserved in a nuclear reaction, therefore, we get

$$Mxc^2 + (E_x + m_xc^2) = (E_Y + M_Yc^2) + (E_y + m_yc^2 \qquad ...(2)$$

where m_x, m_y, M_Y. My all represent respective masses of incident particle, target nucleus, product particle and product nucleus. We now introduce a quantity Q where

$$Q = E_Y + E_y - E_x. \qquad ...(3)$$

From eqns (2) and (3), we have

$$(M_x + m_z - M_Y - m_y)c^2 = Q. \qquad ...(4)$$

The quantity Q is called the energy balance of the reaction or more commonly *Q-value of the reaction.* If this is +ve the reaction is said to be *exoergic.* This occurs if sum of the masses of incident particle and target nucleus is greater than that of masses of the product nuclei. The K.E. of the product nuclei being greater than that of the incident particle. If Q is – ve the reaction is said to be *endoergic, i.e.,* energy must be supplied usually as a K. E. of the incident particle. The term E_Y in eqn. (2) represents the recoil energy of the product nucleus. It is usually small and hard to measure but can be eliminated by taking into account the conservation of momentum. In an experiment to measure a Q value, the bombarding energy E_x and the energy of the ejected particle E_y at some specified angle θ are measured. Thus by applying the laws of conservation of momentum, we have

$$m_{xyx} = M_Y V_Y \cos\phi + m_y y_y \cos\theta \qquad ...(5)$$

and $M_Y V_Y \sin\phi = m_y y_y \sin\theta, \qquad ...(6)$

where v_x, v_y and V_Y are the velocities of incident particle, ejected particle and of product nucleus respectively.

From eqns (5) and (6) eliminating ϕ, we have

$$M_Y{}^2 V_Y{}^2 = mM_Y^2 Y_Y^2 = m_x^2 + m_y^2 y_y^2 - 2m_x m_y v_x v_y \cos\theta. \qquad ...(7)$$

Since $E_x = \frac{1}{2} m_x v_x^2,\ E_y = \frac{1}{2} m_y v_y^2$ and $E_Y = \frac{1}{2} m_Y V_Y^2$

Hence after eliminating v_x, v_y and V_Y, we get

$$2E_Y M_Y = 2E_x m_x + 2E_y m_y - 4(m_x m_y E_x E_y)^{1/2} \cos\theta$$

or $$E_Y = E_x . \frac{m_x}{M_Y} + E_y \frac{m_y}{M_Y} - \frac{2}{M_Y}\left(m_x m_y E_x E_y\right)^{1/2} \cos\theta \qquad ...(8)$$

Substituting the value of E_Y in equation (3), we get

$$Q = E_x \frac{m_x - M_Y}{M_Y} + E_y \frac{m_y + M_Y}{M_Y} - \frac{2}{M_y}\left(m_x m_y E_x E_y\right)^{1/2} \cos\theta \qquad ...(9)$$

This is known as Q-value reaction. It gives the desired relation between the energy released and the measured quantities E_x, E_y and θ. For a special case when we are observing the outcoming particle y at 90° to a collimated beam of projectile, the above relation reduces to

$$Q = E_y\left(\frac{1 + m_y}{M_Y}\right) - E_x\left(\frac{1 - m_x}{M_Y}\right). \qquad ...(10)$$

With the modern techniques, it is possible to measure Q values, for nuclear reaction producing charged particles, to an accuracy of 1 part in a thousand or better.

General Solution of the Q-equation

It is clear from eqn. (9) that the variation of E_y with E_x, for a fixed Q, can also be shown by regarding the Q equation as a quadratic in $\sqrt{E_y}$. Thus its solution can be put in the form

$$\sqrt{E_y} = u \pm \sqrt{\left(u^2 + v\right)}, \qquad ...(11)$$

where $u = \dfrac{\sqrt{\left(m_x m_y E_x\right)}}{M_Y + m_y} \cos\theta$ and $v = \dfrac{M_Y Q + E_x\left(M_Y - m_x\right)}{M_Y + m_y}$.

The energetically possible reactions are those for which $\sqrt{E_y}$ is real and positive. Nuclear reactions are divided energetically as follows

1. Exoergic Reactions

These reactions are possible even for $E_x = 0$. Thus for $E_x \to 0$, equation (11) gives

$$E_y = Q\frac{M_Y}{\left(m_y + M_Y\right)}(Q > 0). \qquad ...(12)$$

The kinetic energy E_y is the same for all angles θ. When $E_x \to 0$.

$$Q = E_y + E_Y \text{ and } \theta + \phi = 180°.$$

Commonly projectile x is lighter than the product nucleus Y. Thus ν is +ve for all values of the bombarding energy and only one of the two solutions of eqn., (11) will be positive. Thus E_y to be single valued $Q > 0$ and $M_y > m_x$. E_y depends on cos θ and is smallest in the backward direction θ = 180°.

2. Endoergic Reactions

All endoergic reactions have negative Q-values. When $E_x \to 0$, eqn. (11) gives $u^2 + \nu$ = –ve and hence $\sqrt{E_y}$ is imaginary. It means that these reactions are not possible. The smallest value of E_x at which reactions can take place is called the *threshold energy*. The reaction first becomes possible when E_x is large enough to make $u^2 + \nu = 0$, hence

$$E_x = -Q\left[\frac{m_y + M_Y}{m_y + M_Y - m_x - \left(\dfrac{m_x m_y}{M_Y}\right)\sin^2\theta}\right].$$

At $\theta = 0$, E_x is minimum and is the threshold energy.

$$\therefore \qquad (E_x)_{th} = -Q\frac{(M_Y + m_y)}{(M_Y + m_y - m_x)} \qquad ...(13)$$

By using relation (4) and an excellent approximation $M_x \geq \frac{Q}{c^2}$, we have

$$(E_x)_{th} = -Q\frac{(m_x + M_X)}{M_X} \qquad ...(14)$$

Thus we see that at the threshold or the reaction particles first appear in the $\theta = 0$ direction with the K.E.

$$E_y = u^2 = (E_x)_{th}\left(\frac{m_x m_y}{m_y + M_Y}\right)^2. \qquad ...(15)$$

As the bombarding energy is raised, particle y begins to appear at $\theta > 0$.

CHARGED PARTICLE REACTION SPECTROSCOPY

The scintillation counter has proved very much useful when too high resolving power is not required and is specially used when coincidences are being studied between the charged particles and γ-rays de-exciting the state formed in the reaction.

The scintillator can be mounted quite close to the target to yield a large solid angle of acceptance and to help to cut down the background pulses from γ-rays, neutrons and cosmic rays. It should be very thin because the charged particles in low energy nuclear spectroscopy have a relatively small range in solid materials. The scintillation spectrometers attain the resolving power $E/\Delta E = 20$ to 30.

For high resolution work, the magnetic spectroscopes are used in nuclear reaction studies. Single focusing instruments, as used in α-spectroscopy, produce a line image of a point source located in the median plane of the magnet, at some distance from the exit of the magnet ⊥ to the median plane because there are no focusing forces acting perpendicular to the median plane.

Point image of a point source is produced by double focusing instruments, can be designed in various ways. A very popular spectrometer of this kind utilizes a magnet with an in homogeneous field similar to the instrument used β-spectroscopy except that it is a sector magnet with both the source and the detector outside the field. We shall describe now a complex instrument used in nuclear reaction studies in association with an 8 MeV Van de Graaff machine, in more detail. Actually in this spectrograph twenty four instruments are placed in one large vacuum enclosure. Each spectograph records a broad range spectrum of particles emitted at a given angle θ from the incident beam. In steps of 7.5° the range of angles is 0° to 172.5°.

The toroidal magnetic circuit comprises forty eight 7.5° cast iron wedges, some of which are thinner than others so that gaps are left between them. These are rigidly clamped together with stainless-steel spacers defining the gaps. Twenty five of the gaps are 3/8" wide, and one gap is 1.5" wide. Each of the former gaps is magnetized by five turns on each of the two pole pieces adjacent to the gap.

The latter gap is magnetized by twenty turns on each side of the gap. The beam reaches the target (placed in the centre of the torus) region by passing through a field free hole through the iron. Lower part of the fig 2.2 shows a cross-sectional side view of the spectrograph with the beam entering hole, the target and the particle trajectories in the 0° gap. The pole face geometry and focal surface geometry are identical for all gaps. The range of energy covered in one exposure is approximately 2.3 : 1, *i.e.,* if 11.5 MeV protons are focussed to the top end of the focal surface, 5 MeV protons will he focussed to the lower end.

After passing through the energy defining slit of the beam analyzing magnet, the beam passes through an electrostatic quadrupole lens designed to produce a reduced image of the energy defining slit on the target. This lens is situated inside a low-carbon steel shielding tube in the 3" diameter hole in the iron torus. It has magnification 1/3 in the vertical direction and unity in the horizontal direction. The target chamber consists of a low-carbon steel pipe of 24" diameter, closed at the bottom by a stainless steel plate used as a reference surface and at the top by a liquid nitrogen cooled copper plate. The beam, after passing through the target, can be collected by a Faraday cup at the opposite end of the target chamber.

The reaction products are recorded on nuclear track plates mounted in rigid holders above each gap. The plates, after exposure, are developed and scanned under microscopes in half millimeter strips across the plate. The number of tracks having correct length and direction are counted.

The number of 1/2 mm strips in 30" is about 1500. This quantity when multiplied by the number of angles normally utilized (24) gives 36,000 data points obtainable with this instrument in one single exposure. The over all resolving power of such an instrument is not determined by the instrument alone but, in general, by the following factors:

(a) the energy stability of the beam of the Van de Graaff generator,

(b) the vertical dimension of the beam spot on the target,

(c) the width of the strips scanned with the microscope across the nuclear track plates,

(d) the effective target thickness and

(e) the aberration of the spectrograph. The resolving power of the above described apparatus was $E/\Delta E \sim 10^3$.

NEUTRON SPECTROSCOPY

Thick photographic emulsion is the simplest example of a neutron spectrometer. Neutrons passing through the emulsion may collide with protons in it and give up a fraction or practically all of their kinetic energy. The energy of the proton is determined by measuring the length of the track, and the neutron energy can then be computed as $T_n = T_p (\cos\theta)^{-1}$. Mere θ is the angle between the directions of the proton and the incident neutron. The direction of the entering neutron is presumably known and the direction of the proton or the track in the emulsion is measured under the microscope. The resolving power is limited by the straggling of the protons in the emulsion.

Scintillators containing hydrogen can be used to detect the protons recoiling from n-p collisions in the scintillator material. Unless the scintillator is very large, the total energy of the neutron is not given to the scintillator. Since it is very difficult to measure the angle between the neutron direction and the proton direction, hence the scintillation counter is not very well suited for neutron spectroscopy. Solid state detectors have recently come into use in neutron spectroscopy in the so-called *sandwich counters*.

A thin film containing Li^6 is sandwiched between two solid state counters. In passing through the sandwich neutrons (1 to 10 MeV) will have a reasonably high probability of initiating the Li^6 (n, t) He^4. If the alpha particle enters one of the solid state counters and the triton enters the other, a coincidence can be registered and the sum of the pulse heights from these counters is a measure of the neutron energy. The detection efficiency of sandwich counters is of the order of 10^{-6} or 10^{-7}.

With pulsed accelerators and devices to measure the time of flight of neutrons over a path from target to detector, the most exact work in neutron spectroscopy has been performed. The energies are determined through a measurement of velocities of neutrons. The accelerator may be a Van de Graaff generator and the detector may be a scintillation counter with associated electronic equipment designed specially 10 measure the time delay (the time in which a pulse strikes the target) spectrum. The flight time for neutrons over a path length L in metres depends on the kinetic energy T (in MeV) of the neutron as

$$t = \frac{L}{v} = L\sqrt{\frac{m}{2T}} = \frac{L}{c}\sqrt{\frac{mc^2}{2T}} = \frac{72.3L}{\sqrt{T}} \text{ n sec.} \quad ...(1)$$

This is quite good for neutron spectroscopy but decreases with the increase of neutron energy because time of flight decreases with the K.E. of the neutron while Δt remains constant. The resolving power can be increased by increasing the time of Eight by making the path longer.

The time of flight spectrum of neutrons from a Ne^{20} gas target bombarded with 3.07 MeV deuterons. The intense peak to the far right corresponds to the γ-rays arriving first ($v = c$). The target to counter distance L was 2.02 m. The time difference between channels was approximately 2.3 nsec with zero time at about channels 102.

NUCLEAR CROSS SECTION

It is usually expressed in terms of an effective area presented by a nucleus towards the beam of bombarding particles, such that the number of incident particles that would strike such an area, calculated upon a purely geometrical basis, is the number observed to lead to the reaction in question. This effective area is called the cross-section for that reaction. Thus the probability of occurrence of a particular nuclear reaction is described by the effective cross-section for that process. The cross-section may also be denned as :

1. The probability that an event may occur when a single nucleus is exposed to a beam of particles of total flux one particle per unit area.

2. The probability that an event may occur when a single particle is shot perpendicularly at a target consisting of one particle per unit area.

Partial Cross-section

The total nuclear cross-section is the effective area possessed by a nucleus for removing the incident particles from a collimated beam by all possible processes. This can be written as a sum of several partial cross-sections which represent the contributions of the various distinct, independent processes which can remove particles from the incident beam. Thus we have

$$\sigma_t = \sigma_s + \sigma_r, \quad ...(1)$$

where σ_l is the total cross-section, as the cross-section that produces measurable scattering and σ_r, the reaction cross-section. Scattering cross-section can be classified as : inelastic scattering and elastic scattering processes, Thus, we get :

$$\sigma_s = \sigma_{cl} + \sigma_{incl.} \quad ...(2)$$

These partial cross-sections can still be sub-divided. In the case of elastic cross-section separate partial cross-sections cannot be written because of the possibility of interference between them. On the other hand all inelastic scattering processes are incoherent and their cross-sections are additive,

$$\sigma_{incl.} = \sigma_s + \sigma_a + \sigma_n + ... \quad ...(3)$$

Similarly or the partial cross-section for capture may be sub-divided into component parts associated with the various possible nuclear reactions. Hence we can write

$$\sigma_r = \sigma' + \sigma'' + \sigma''' + ... \quad ...(4)$$

Differential Cross-section

The cross-section which defines a distribution of emitted particles with respect to the solid angle is called differential cross-section. It is defined as $d\sigma/d\Omega$ and the partial cross-section for a given process is given-by as follows

$$s = \int \frac{d\sigma}{d\Omega} d\Omega. \quad ...(5)$$

Determination of Cross-section

Experimental values of a can be obtained by the measurement of the attenuation of an incident beam in a known thickness of material. If a monoenergetic beam consisting of *l* particles per unit time distributed uniformly over an area A is allowed to pass through a thin sheet of target material of thickness Δx and having n nuclei per unit volume then we have the available nuclear target area which is the product of the cross-section per target nucleus (σ) and the number of target nuclei lying with in the area A.

$\therefore$ Probability for one incoming particle to hit a nuclear target = Nuclear target area $\sigma nA\ \Delta x$/Total target area A. If the nuclear reaction produces N light product particles per unit time, then the reaction probability is defined as the ratio N/I. Hence we have

$$\frac{N}{I} = \frac{\sigma n A\, \Delta x}{A} \text{ or } \sigma = \frac{N}{\left(\frac{I}{A}\right)(nA\Delta x)}. \qquad ...(6)$$

Hence the nuclear cross section can be defined as the number of light product particles per unit time per unit incident flux and per target nucleus. If a bombarding particle is producing a variety of light reaction products N_1, N_2,... per unit time, then the total cross-section is defined as

$$\sigma_{total} = \frac{N_1 + N_2 + ...}{\left(\frac{I}{A}\right)(nA\, \Delta x) \cdots} \qquad ...(7)$$

If the incident flux (beam intensity) is I_0 and the flux after penetrating a distance x is I, it will become I – dI after a further penetration of distance dx. This reduction in flux is caused by the various collision events occurred with in this distance dx.

$$\therefore \qquad -\,dI = N = n\sigma\, dxI \text{ or } -\frac{dI}{I} = \sigma n\, dx. \qquad ...(8)$$

This expression contains the product of σ, the cross-section per nucleus or the *microscopic cross-section* and n, the number of nuclei per unit volume. The product is called the *macroscopic cross-section* and is denoted by Σ. Thus we have

$$-\frac{dI}{I} = \Sigma dx \quad \text{or} \quad I = I_0 e^{-\Sigma_x} \qquad ...(9)$$

The experimental arrangement consists of a source and a detector between which can be placed a slab of the experimental material. If I, is the particle intensity reaching the detector in the absence of the absorbing material and I is the value when the slab of thickness x, containing n atoms per unit volume is present, the total cross-section o can be calculated. In many nuclear reactions the light product particles are not produced in an isotropic manner. We thus define differential cross-section $d\sigma/d\Omega$ in terms of the number of light products dN emitted per unit time in a small solid angle $d\Omega$ at some angle θ with the direction of incident beam. From eqn. (22), we have

$$\frac{1}{I} = \frac{dN}{d\Omega} = \frac{nA\,\Delta x\left(\dfrac{d\sigma}{d\Omega}\right)}{A} \quad \text{or} \quad \frac{d\sigma}{d\Omega} = \frac{\dfrac{dN}{d\Omega}}{\left(\dfrac{1}{A}\right)(nA\,\Delta x)}.$$

Macroscopic Cross-section and Mean Free Path

The equation (9) shows that Σ corresponds to linear absorption coefficient and has the dimensions of reciprocal length. The penetration distance λ can be defined as the path length over which the neutrons can travel without suffering a collision with a target. This distance is known as *mean free path.* Since beam intensity is proportional to the beam particle density q, hence we can write eqn. (9) as

$$q = q_0 e^{-\Sigma x}.$$

$$\text{The average path length } \lambda = \frac{\text{Total path length}}{\text{Total no. of particles}} = \frac{\int_0^{\infty} x\,dq}{q_0}$$

$$= \frac{1}{\Sigma}\int_0^{\infty}(-\Sigma x)e^{-\Sigma x}\,d(\Sigma x) = \frac{1}{\Sigma} \qquad \text{...(10)}$$

Thus the average path length λ between successive collisions is equal to the reciprocal of the macroscopic cross-section Σ.

SOLVED EXAMPLES

Example 1:

When F^{19} is bombarded with protons a (p, n) reaction with subsequent α-emission occurs. Calculate the excitation energy of the compound nucleus that corresponds to the resonance with a proton energy of 4.99 MeV.

$$_1H^1 + {}_9F^{19} \rightarrow ({}_{10}Ne^{10})^* \rightarrow {}_{10}Ne^{19} + {}_0n^1.$$

The convertible energy of the protons or the energy of relative motion is

$$E = \frac{E_0}{1+\frac{m}{M}} = \frac{4.99}{1+\frac{1}{19}} = 4.745 \text{ MeV.}$$

This binding energy contribution of the absorbed proton in the compound nucleus is

$$\Delta E = [M({}_9F^{19}) + M({}_1H^1) - M({}_{10}Ne^{28})]mu$$

$$= [18.998405 + 1.007276 - 19.992440]mu$$

$$= 0.013241 \text{ mu} = 12.332 \text{ MeV}$$

$\therefore$ Excitation energy of the compound nucleus

$$= 4.745 + 12.332 = 17.077 \text{ MeV.}$$

Example 2:

A thin sheet of Co^{59}, 0.04 cm, thick, is irradiated with a neutron beam of flux density 10^{12} neutrons per cm^2 per sec for a period of 3 hr. If the cross-section for neutron capture by Co^{59} is 30 barns, calculate the number of nuclei of the isotope Co^{50} produced at the end of the irradiation period per cm^2 and the initial β-activity of the sample. Given: half-life of Co^{60} 5.2 years and density of Co^{59} 8.9 gm/cm^3.

Solution:

$$\text{Number of reactions/second/m}^2 = N_0\sigma It = N\left(\frac{\rho}{M}\right)\sigma It$$

$$= 6.02252 \times 10^{26} \times \left(8.9 \times \frac{10^3}{59}\right) \times 30 \times 10^{-28} \times 10^{16} \times 0.04 \times 10^{-2}$$

$$= 0.092 \times 10^{15}.$$

$\therefore$ No. of Co^{60} nuclei produced = No. of transmutation in 3 hours

$$= 1.092 \times 10^{15} \times 3 \times 60 \times 60 = 118 \times 10^{17}.$$

Initial activity per $m^2 = \lambda \times 118 \times 10^{17}$

$$= \frac{0.693 \times 118 \times 10^{17}}{5.2 \times 365 \times 24 \times 60 \times 60}$$

$$= 5 \times 10^{6} \text{ decays/sec} = 134 \text{ micro curies.}$$

Example 3(a):

The total cross-section of nickel for 1 MeV neutrons is 3.5 barns. What is the fractional attenuation of a beam of such neutrons on passing through a sheet of nickel 0.01 cm in thickness? Given, density of nickel 8.9 gm/cm³.

Solution:

Macroscopic cross-section $\Sigma = N_0\, \sigma = \left(\frac{\sigma}{M}\right) N\sigma$

$$= \left(\frac{8.9 \times 10^{3}}{58}\right) \times 6.02252 \times 10^{26} \times 3.5 \times 10^{-28}$$

$$= 32.37 \text{ m}^{-1}$$

We know the relation $\sigma I_0 e^{-\Sigma x}$ or $2.3026 \log_{10} \frac{I_0}{I} = \Sigma x$

$$\therefore \qquad \frac{I_0}{I} = \text{Anti log} \frac{32.37 \times 0.01 \times 10^{-2}}{2.3026} = 0.0014.$$

Example 3(b):

The neutron capture reaction of Au^{197} at neutron energies upto a few hundred eV is characterized by a number of resonances. The most prominent is at 4.906 eV and has $\Gamma_\gamma = 0.124$ eV and $\Gamma_n = 0.007\, E^{1/2}$ eV. The compound nucleus formed in this resonance absorption has spin 2. Calculate the peak cross-section.

Solution:

$\Gamma_n = 0.007\, E^{1/2} = 0.007 \times (4.906)^{1/2} = 0.0155$ eV.

We know the relation for resonance cross-section

$$\sigma(n, \gamma) = \frac{\lambda^2}{\pi} \frac{2I_c + 1}{2(2I+1)} \frac{\Gamma_\gamma \Gamma_n}{(\Gamma_\gamma + \Gamma_n)^2} = \frac{h^2}{2\pi m E} \frac{2I_c + 1}{2(2I+1)} \frac{\Gamma_\gamma \Gamma_n}{(\Gamma_\gamma + \Gamma_n)}$$

$$= \frac{\left(6.625\times10^{-34}\right)^2}{2\times3.14\times1.67\times10^{-27}\times4.906\times1.6\times10^{-15}}\times\frac{2(2+I)}{2(3+1)}$$

$$\times\frac{0.124\times0.0155}{(0.1395)^2}$$

$= 3.446 \times 10^4$ barns.

Example 3(c):

Estimate the relative probabilities of (n, n) and (n, γ) in indium, known to have a neutron resonance at 1.44 eV with a Γ of 0.1 eV and cross-section of 28000 barns.

Solution:

At resonance ($E = E_0$) Breit-Wigner formula simplifies to

$$\sigma(n, \gamma) = \frac{\lambda^2 \Gamma_n \Gamma_\gamma}{4\Gamma^1}.$$

The de Brogile wavelength $\lambda = \dfrac{h}{\sqrt{(2mE)}} = 2.4 \times 10^{-11}$ m.

$$\Gamma_n\Gamma_\gamma = \frac{4\sigma\Gamma^2}{\lambda^2} = \frac{4\times2.8\times10^{-24}\times(0.1)^2}{\left(2.4\times10^{-11}\right)^2}$$

$$= 1.5 \times 10^{-4}$$

Since $\Gamma_\gamma \simeq \Gamma = 0.1$, hence $\dfrac{\Gamma_n}{\Gamma_\gamma} = 0.015$.

Example 4:

Estimate by the aid of the semi-classical stripping theory the angles at which the (d, p) differential cross-section has its maximum for the following l-values : $l = 0, 1, 2, 3, 4$. Use $R = 5f$, $E_d = 10$ MeV and $E = 14$ MeV.

Solution:

From semi-classical theory for stripping reaction, the angles are given by the reaction

$$\cos\theta \le \frac{\left(k_d^2 + k_p^2\right)}{2k_d^2} - \frac{l^2}{2k_d k_p R^2},$$

where $k_d = \sqrt{\frac{(2m_d E)}{\hbar}} = 0.98\ f^{-1}$ and $k_p = \sqrt{\frac{(2m_p E)}{\hbar}} = 0.82\ f^{-1}$.

$$\therefore \quad \cos\theta \le \frac{0.98^2 + 0.82^2}{2\times0.98\times0.82} - \frac{l^2}{2\times0.98\times0.82\times5^2}$$

$$\le 1.015 - \frac{l^2}{40.18}.$$

Hence from natural cosines table, we get

l	0	1	2	3	4
θ_{max}	0	90°	24°	38°	52°

Example 5:

Calculate the Q-value for the formation of P^{30} in the ground state in the reaction Si^{29} (d, n) P^{30} from the following cycle of nuclear reactions:

(i) $P^{31} + \gamma \rightarrow P^{30} + n - 12.37$ MeV

(ii) $P^{31} + p \rightarrow Si^{28} + He^4 + 1.909$ MeV

(iii) $Si^{28} + d \rightarrow Si^{29} + p + 6.246$ MeV

(iv) $2d \rightarrow He^4 + 23.834$ MeV

Solution:

Reaction (i) can be written as

$P^{30} + n \rightarrow P^{31} + g + 12.37$ MeV. ...(ii)

Formation of P^{30} is given by the reaction

$Si^{29} + d \rightarrow P^{30} + n + Q.$

Combination of these equations with reactions (ii), (iii) and (iv) gives

$4d = 2He^4 + Q + 44.359$ MeV

$Q = 2(2d - He^4) - 44.359$

$= 47.668 - 44.359 = 3.309$ MeV.

Example 6:

A tritium gas target is bombarded with a beam of monoenergetic protons of kinetic energy 3 MeV. What is the K.E. of the neutrons emitted at 30° to the incident beam? Atomic masses are : H^1 1.007276 mu, n^1 = 1.008665 mu, $_1H^3$ = 3.016056 mu, and He^2 = 3.016030 mu.

Solution:

Nuclear reaction in question can be written as

$$_1p^1 + {}_1H^3 \rightarrow {}_2He^3 + {}_0n^1 + Q,$$

where $Q = m(_1p^1) + m(_1H^3) - m(_2He^3) - m(_0n^1)$

$= 1.007276 + 3.016050 - 3.016030 - 1.008665$

$= -0.001369$ mu $= -1.2745$ MeV.

The K.E. associated with the emitted neutron is given by eqn. (11) as

$$\sqrt{E_n} = u \pm \left(u^2 + v\right)^{1/2},$$

$$\text{where } u = \frac{\left(m_p m_n E_p\right)^{1/2}}{mH_e + m_n}\cos\theta = \frac{(1.007276 \times 1.008665 \times 3)^{1/2}}{3.016030 + 1.008665} \times \frac{\sqrt{3}}{2}$$

$= 0.3753$

$$\text{and } v = \frac{m_{He}Q + E_p\left(m_{He} - m_p\right)}{m_{He} + m_n}$$

$$= \frac{-3.016030 \times 1.2745 + 3 \times 2.008754}{3.016030 + 1.008665} = 0.5424$$

$\therefore \sqrt{E_n} = 0.3753 \pm 0.8266 = 1.2019$

or E_n = 1.444 MeV.

EXERCISES

1. Calculate the threshold energy required to initiate and the reaction P^{31} (n, p) Si^{31}, given M_p = 1.00814, M_n = 1.00898, M_P = 30,98356 & M_{Si} = 30.98515.

Calculate also the maximum energy of β-decay of Si^{31} to P^{31}.

2. In the B^{10} (α, p) C^{13} reaction with 4.77 MeV α-particles the two most energetic proton groups which were observed to be emitted at the angle of 90° with the direction of the incident α-particle beam had energies of 6.84 MeV and 3.98 MeV respectively. Give the information about the energy levels of the residual nucleus.

3. The Q-valve of the reaction $Ra^{224} \rightarrow Em^{222} + He^4$ is 4.88 MeV. The radium nucleus is originally at rest. Calculate the K.E. of each disintegration product.

4. In the following deuterium reactions the reaction energy is as stated;

   N^{14}(d, p) N^{15}, Q = 8.61 MeV; N^{15} (d, α) C^{13}, Q = 7.68 MeV;

   C^{13} (d, α) B^{11}, Q = 5.16 MeV; B^{11} (α, n) N^{14}, Q = ?;

   Calculate Q value of the fourth reaction. Given : $M\alpha = 4.003873$,

   $M_d = 2.014740$, $M_p = 1.008145$ and $M_n = 1.008996$.

5. Assuming a resonance cross-section 5.5×10^4 barns for Cd^{113} for Cd^{113} at a resonance energy of 0.176 eV. Calculate the relative probability of a neutron emission as compare to a resonance capture if $\Gamma = 0.113$ eV and $\Gamma_\gamma << \Gamma_n$.

6. Calculate and compare the collision and absorption mean free paths for neutrons in graphite. Given : $\sigma_s = 4.8$ barns, $\sigma_a = 3.2 \times 10^{-3}$ barns and $\rho = 2.25$ gm/cm³.

7. An indium foil of 2 cm² cross-section and 10^{-3} cm, thickness is exposed to a spread beam of neutrons of uniform energy. Calculate the number of neutron capture that will occur during a 3 min exposure of the foil. Given; $I = 5 \times 10$ neutrons/cm² sec. and $\sigma_u = 190$ barns.

8. The B^{10} (α, p) C^{13} reaction shows among other a reasonance for an excitation energy of the compound nucleus of 13.23 MeV. Calculate the mean life of the nucleus for this excitation if level width is 130 keV.

4

Nuclear Detection

INTRODUCTION

Neutrons are uncharged particles, therefore, produce very little direct ionization about one ion pair per metre of path in passing through matter. They cannot be detected in any instrument whose action depends on the ionization due to the entering particle. For reasons of practical efficiency, the secondary particles must be generated at energies which can produce ionization conveniently detectable. The secondary charged particles may be

(a) the protons released by collisions of neutrons with hydrogen nuclei,

(b) the direct result of nuclear disintegration produced by neutrons,

(c) the radioactive disintegrations from product nuclei which become radioactive as a result of neutron capture or

(d) the strongly ionizing fission fragments produced by the absorption of neutrons in fission reactions.

Two general principles have been employed in the detection of neutrons, first, use is made of the charged particles produced by the interaction of neutrons with various substances introduced into the counter, and second, advantage is taken of the recoil of nuclei of light elements after being struck by neutrons. For the detection of slow neutrons the first method is used while for fast neutrons we use second method.

1. Slow neutron detection

(a) Boron detectors

Detection process is based on the reaction.

$$_5B^{10} + {}_0n^1 \rightarrow {}_3Li^7 + {}_1He^4 + 2.78 \text{ MeV}.$$

For thermal neutrons (0.025 eV) the cross-section for this reaction is 3770 barns. The 2.78 MeV reaction energy being carried off by the resulting lithium nucleus and alpha particles which produce considerable ionization in their tracks. However, if the target is natural boron the effective cross-section is reduced to 710 barns as the abundance of the B^{10} isotope is only 18 83%. Better results are obtained if the boron compound is enriched in B^{10} isotope. The cross-section for the above reaction follows the 1/v law and falls off with increasing neutron energy. Hence these counters are useful for slow neutrons. Ionization chambers with boron coated electrodes or proportional counters filled with boron trifluoride vapour (enriched in B^{10}) are in common use. Full advantage of the proportional counter is taken to discriminate neutron induces pulses from smaller background gamma pulses.

The BF_s proportional counter tube has been modified by Hanson and Mc Kibben (1947) to make it sensitive to fast neutrons in the energy range 10 keV to 3 MeV. This counter (shown in Fig. 4.7) is called *long counter*. The long (10") thin (diameter 1/2") cylindrical BF_3 proportional counter tube is embedded in a cylinder of paraffin (dia 8"). Around this paraffin is a layer of an absorber for slow neutrons which is covered with an outer layer of paraffin. Fast neutrons are moderated (*i.e.,* they are rapidly reduced to thermal energies by elastic collisions with the nuclei of the moderator) in the paraffin wax. The neutrons incident at an energy of about 100 keV have a slightly better chance of detecting than those at lower energy. Experimental tests have verified the expectation that the long counter has an efficiency for the detection of neutrons which is practically constant over a considerable range of neutron energies. The detector efficiency for neutrons incident on the 8 inch circle on the front face is less than 1 per cent.

(b) Fission Chamber

The relatively high cross-sections for fission by thermal neutrons of U^{235}, Np^{237} and Pu^{239} has led to the development of thermal neutron detectors. The fission fragments have large energies (> 50 MeV) and if one of them escapes from the thin surface layer of the fissionable material into a gas of the chamber, a large, easily distinguished, ionization pulse is generated. A simple type of fission chamber for observing neutrons thus consists of an ionization chamber for which one electrode is coated with uranium oxide, preferably enriched in the uranium 235 isotope. There are following limitations on the design of fission counters :

(i) One is imposed by strong a-particle background-always present in fission chambers.

(ii) Another involves the short range of the fission fragments.

To have the pulses produced by the fission fragments much larger than the pulses from the α-particles, the sensitive layer of fissionable material introduced into the counting chamber must be quite thin (~ 100 micrograms/cm^2), also the resolution of the recording system must be high to reduce the effect of coincident α-particle pulses.

(c) Activation Method

The important discovery made by Fermi, that many elements exposed to a flux of the thermal neutrons become radioactive, has led to a useful method of detecting allow neutrons. The feasibility of detecting the induced radioactivity depends on its life time, which cannot be appreciably shorter than the time which must elapse between exposure to neutrons and measurement of the induced activity, on the other hand, the life time must not be so long that the decay rate is negligible. A number of elements have a large activation cross-section for the (n, γ) reaction. Thin foils of a suitable element (manganese, gold, indium and rhodium are often used) are exposed to a neutron flux for a time interval greater than six or seven times the half life of the induced activity. They are then removed and the extent of the induced activity is determined by counting the emitted radiations with an appropriate Geiger counter, ionization chamber, scintillation counter or other radiation detector. Indium has a large cross-section for the reaction at low energy range, therefore, is a very efficient detector for slow neutrons at the energy. One method of estimating the thermal component of a neutron flux is to measure the induced activity in a bare indium foil and other wrapped in cadmium foil (~ 1 mm. thick). For low energy neutrons (0.3 eV} cadmium is nearly opaque (the cross-section for the reaction is several thousand barns but it does not yield a radioactive isotope) whereas for epithermal neutrons the Cd foil is nearly transparent.

Szilard and Chalmers in 1934 devised a technique for separating radio-nuclei, induced by n, γ reaction, from their isotopic environment. The γ-rays emitted from product atoms in the (n, γ) reaction can provide recoil energy E_γ sufficient to break the chemical bond between the radioactive atoms and thereby to change the chemical state of the product nucleus as compared to the normal nucleus in the medium. The recoil energy in eV from the emission of a γ-ray of energy E_γ in MeV is given as

$$E_r = 536 \frac{E_r^2}{A} \quad \text{...(1)}$$

where A is the mass number of the recoiling atom. For $E_\gamma = 7.5$ MeV and A = 100, the recoil energy is 300 eV which is fairly large than necessary to disrupt chemical bonds (1 – 5 eV). The above reaction known as *Szilard-Chalmers* reaction is used to obtain radioactive samples of high specific activity, especially when only relatively weak neutron source were available.

(d) Scintillation Counter

Considerable work has been done by various investigators to develop scintillation counters for slow neutrons. Because neutrons cannot produce scintillation directly, the methods used are based on the addition to the scintillator material of an element which interacts with neutrons to yield particles capable of producing scintillations directly. One of the early methods involved the preparation of alternate thin layers of a boron compound and ZnS. The α-particles released by the neutrons from the boron were detected by the well known ability of activated ZnS. The α-particles were relatively insensitive to γ-rays, they did not have a very satisfactory efficiency for the detection of neutrons. Not all ionizing particles from the disintegration of the boron succeeded in producing scintillations and only a fraction of the light from the scintillations reached the photo-multiplier, much of the light being lost by reflection. Both these difficulties can be corrected to some extent by fusing the ZnS in a glassy layer of boric anhydride. The resulting water is applied directly to the window of the photomultiplier tube. When large volume detectors are required, liquid scintillators containing boron compounds are often used. Methyl borate dissolved in terphenyl and phenyl-cyclohexane (main solvent) is an efficient detector of slow neutrons.

2. Intermediate Neutron Detection

Cross-sections m this range of neutron energies are generally so low that to yield adequate intensities of ionizing radiations thick layers of the sensitive elements are required. Sensitive elements which emit γ-rays are not subject to the same limitation. Rae and Bowey (1953) have designed a detector which utilizes the soft gamma rays from the reaction

$$_5B^{10}\ (n, \alpha)\ _3Li^{7*} \rightarrow {_3Li^7} + \gamma(480\ \text{keV})$$

for measuring neutrons of intermediate energies. Layers of $_5B^{10}$ which are opaque to neutrons with energies upto about 1 keV may then be used as the detector. By using a NaI phosphor with a rapid rate of rise of the luminous response in a single, channel scintillation spectrometer, the disturbing background from other radioactive sources and the cosmic radiation was materially reduced. In addition a thorough system of lead screening and neutron shields were provided. The detector consists of two photomultipliers, each with NaI phosphor, faced the boron detector from opposite sides. As the amount of boron used is no longer opaque to neutrons above 1 keV, a drop of ~ 15% in the efficiency is expected at 5 keV and of ~ l% at 50 keV.

3. Fast Neutron Detection

(a) Proton-recoil Detector

This method for detecting fast neutrons is based on the observation of the ionization produced by the recoiling protons in the elastic scattering of neutrons by hydrogeneous materials. For this purpose a proportional counter may be filled with hydrogen or deuterium, but a better procedure is to use argon, or one of the heavier inert gases as a filling gas and to place a thin sheet of a hydrogeneous material such as paraffin at one end of the chamber. Fast neutrons striking the target (paraffin) cause the ejection of protons, which produce ionization in their paths through the counter and so can be detected. If neutron's energy is not large, the recoil pulses will be masked by the small pulses caused by γ-rays or cosmic rays. In 1954 Sun and Richardson designed a multiple-wire proportional counter contained 1.2 atmospheres of methane and operated at 3400 volts. This has beer found to be almost completely insensitive to γ-radiations when operated at an over all efficiency of 0.17 percent for detecting neutrons of energies ranging from 0.3 to 10 MeV.

(b) Fission Process

The isotope U^{235} has a very large cross section for slow neutrons, where as U^{288} has a large cross section for fast neutrons. Hence a fission chamber for the detection of fast neutrons can be designed using a uranium compound which has been depleted in the lighter isotope. This device is useful for fast neutron detection only. For both slow and fast neutrons a mixture of U^{236} and U^{238} is used. The slow neutron component can again be cut out if desired, by shielding the counter either with cadmium or boron.

(c) Scintillation Counter

Hornyak (1952) designed scintillators which were effective for fast neutrons by molding a scintillating phosphor which was sensitive to recoiling protons, such as ZnS, into a plastic serving as a fairly dense radiator. Another approach to this problem has been suggested by McCrary, Taylor and Bonner (1955). They have found that isolated spheres of anthracene will yield brighter scintillation from recoil protons than from the electrons released by γ-rays if the diameter of the anthracene spheres is in the range from 3 to 8 mm. The same secondary electron with energy of the order of 5 MeV can be prevented from exciting more than one sphere if several millimeters of glass or quartz separate the individual spheres.

Taylor. Lonsjo and Bonner in 1954, used an ingenious scintillation detector of fast neutrons (Fig 4.8). A number of small plastic scintillating spheres is immersed in a non-hydrogeneous fluid ($C_6Cl_3F_3$) of nearly the same refractive index. The liquid is a non-scintillator. It optically couples the plastic spheres via a quartz rod to the 4 cm diameter cathode of a photomultiplier tube. Since γ-rays produce small pulses, the neutron pulses are easily distinguished from γ-pulses. The optimum diameters of the plastic spheres for detecting neutrons of kinetic energy 3.5, 4.5, 7.0 and 14 MeV are 1.5, 2.5, 6.0 and 10 mm respectively.

(d) Photographic process

The nuclear track emulsion is an excellent detector for the protons which recoil from neutrons in the neutron-proton-scattering process when the neutron energy exceeds about 2 MeV. Fast neutrons can be observed by photo graphing the tracks of recoil protons in a hydrogen water vapour filled cloud chamber.

4. Ultra-High-Energy Detection

The efficiencies of detectors of neutrons with energies greater than 50 MeV are usually low. Proton recoils from a thin hydrogeneous target may be measured with telescope consisting of three proportional counter tubes in line. The telescope can be made insensitive to protons generated by neutrons of energies lower than some arbitrary value by introducing suitable absorbers in between two telescopes. The advantage of this telescope is that the total range of the recoil protons can be measured and the energy of the neutrons can be deduced from this measurement. Because neutrons of energies greater than 50 MeV can induce fission in some of the heavy stable elements (Bi), fission chambers made with such elements as electrodes may be used to detect these ultra-high-energy neutrons. Liquid scintillators are also used for the detection of ultra-high-energy neutrons.

5. Threshold Detectors

Numerous reactions of neutrons with nuclei have definite thresholds for the energies of the neutrons below which the reactions do not occur. These thresholds extent from a few tenths of 1 MeV to 20 MeV and higher involve either (n, 2n) or charged particle reactions. For example the reaction ($_6C^{12} + {_0n^1} \rightarrow {_6C^{11}} + 2{_0n^1}$) does not occur unless the neutron energy is greater than 22 MeV, the threshold energy of the reaction. Thus we see that $_6C^{12}$ can be used as a threshold detector for neutrons whose energies exceed 22 MeV.

SLOWING DOWN NEUTRONS

The neutrons emitted in nuclear reactions have an energy of a few MeV. Since neutrons are not electrically charged they cannot loose energy by ionization but are slowed down by successive scattering in matter until they reach in thermal equilibrium with their surroundings. After this they diffuse like gaseous molecules until they are captured by nuclei.

A. Energy Loss

In 1934, Fermi and his colleagues showed that the neutrons are slowed down in the hydrogeneous materials and the slower neutrons have a greater probability of inducing radioactivity than do more energetic neutrons. The amount of energy that a neutron losses in a single collision can be calculated by using the equations of conservation of energy and momentum. Another method, which is very simpler involves the use of two reference systems. The first is the L-system, in which the target nucleus is initially at rest and is approached by the incident neutron. The second is the C. M. system in which the centre of mass of the colliding particles is at rest initially and throughout the collision. *In the L-system,* before the collision, the neutron of mass m moves towards the nucleus of mass M, which is assumed to be at rest, with speed v_0. The total linear momentum of the colliding particles is zero in C. M. system. If v_c be the velocity of incident particle in this system and V that of the target nucleus, then

$$mv_c - MV = 0. \quad ...(1)$$

It is clear that the velocity of the target nucleus in the C.M.-system is equal and opposite to the velocity of the C. M.-system in the L-system. Thus we can write

$$v_0 = v_c + V. \quad ...(2)$$

$$\therefore \quad V = \frac{mv_0}{(m+M)}$$

and

$$v_c = \frac{Mv_0}{(m+M)} \quad ...(3)$$

If the ratio of moderator mass to neutron mass M/m is called A, the mass number of the moderator nucleus, the above relations can be written as

$$V = \frac{v_0}{(1+A)} \text{ and } v_c = v_0 \frac{A}{(1+A)}. \quad ...(4)$$

Principle of conservation of energy follows that the speeds of the colliding particles are not changed after the collision in C. M.-system. If the neutron moves at an angle ϕ with its initial direction after the collision, to conserve total momentum, the nucleus must move off at an angle $(180 + \phi)$ with the direction of the incident neutron.

Thus the total effect in the C.M.-system is a rotation of the particle system by an angle ϕ only. In the L-system, the neutron is scattered through an angle θ, has a velocity v_1 and the nucleus has a velocity v_2. The v_1 is the vector sum of the neutron velocity in C.M.-system and the velocity of C.M. By using vector diagram.

$$v_1^2 = v_c^2 + V^2 + 2V_{vc}\cos\phi = v_0^2 \frac{A^2 + 2A\cos\phi + 1}{(1+A)^2}$$

If E_0 is the neutron energy in the L-system before the collision and E_1 the neutron energy in the same system after the collision. Then we can write

$$\frac{E_1}{E_0} = \frac{v_1^2}{v_0^2} = \frac{A^2 + 2A\cos\phi + 1}{(1+A)^2} \quad ...(5)$$

This relation indicates that $E_1 = E_0$ for a grazing collision ($\phi = 0$), which means that no energy is transferred from the neutron to the moderator nucleus and $E_1 = \alpha E_0$ for a head on collision ($\phi = \pi$), which means that maximum energy is transferred from neutron to the moderator nucleus. Here

$$\alpha = \frac{(A-1)^2}{(A+1)^2}. \quad ...(6)$$

Thus in a single collision maximum possible energy loss by the neutron

$$(\Delta E)_{max} = E_0 - (E_1)_{min} = E_0(1 - \alpha).$$

$$\therefore \text{ Maximum fractional energy loss } \left(\frac{\Delta E}{E_0}\right)_{max} = 1 - \alpha \quad ...(7)$$

Since α depends on A, the maximum loss of K. E. depends on the atomic weight of nucleus encountered. For example, for hydrogen A = 1 and $\alpha = 0$ and the maximum loss of kinetic energy is E_0.

For larger values of A, α can be written as

$$\alpha = \left(\frac{1-1}{A^2}\right)\left(\frac{1+1}{A}\right)^{-2} = \left(\frac{1-2}{A}\right)^2 \simeq \frac{1-4}{A}.$$

$$\text{or Maximum fractional energy loss} = I - \alpha = \frac{4}{A}. \quad ...(8)$$

Thus we see that the light nuclei are more effective moderators than heavy nuclei and the maximum loss of energy is proportional to the initial energy.

B. Scattering Angles in L-system and C. M. System

The scattering angle ϕ in the C.M.-system can be related to the scattering angle θ in the laboratory system as :

$$v_1 \cos \theta = v_c \cos \phi + V. \qquad ...(9)$$

Substituting here values of v_1, v_c and V, we have

$$\cos \theta = \frac{1 + A \cos \phi}{\left(A^2 + 2A \cos \phi + 1\right)^{1/2}}. \qquad ...(10)$$

When the collision is against a very heavy nucleus A >> 1, the second and third term in the denominator of this equation can be neglected compared with A^2. Thus we can write

$$\cos q = \cos \phi + \frac{1}{A}. \qquad ...(11)$$

Thus the scattering angle is same in two systems if A + 1.

C. Average Energy Loss Per Collision

Before we proceed to calculate the average energy that a neutron is going to loss per collision we should remember the experimental fact that in the C.M. system all scattering angles are equally probable. From this empirical law we can now deduce the probability of occurrence of a final kinetic energy E_1. To find the probability of a neutron being scattered through an angle lying between ϕ and $\phi + d\phi$, we describe a unit sphere about the scattering centre O and evaluate the corresponding space angle $d\omega_\phi$, as

$$d\omega_\phi = 2\pi \sin \phi \, d\phi.$$

Since the probability of a particle being scattered in such a way that it passes through area dS is equal to the fraction of the projected area on to the unit sphere as compared to the total surface area of this sphere. Hence.

$$dW = \frac{d\omega_\phi}{4\pi} = \frac{1}{2} \sin \phi \, d\phi$$

$$= -\frac{1}{2} d(\cos \phi). \qquad ...(12)$$

The eqn. (5) can be written in terms of a as

$$\cos\phi = \frac{1}{1-\alpha}\left[\frac{2E}{E_0}-(1+\alpha)\right]$$

$$\therefore \qquad dW = -\frac{dE}{E_0(10-\alpha)}. \qquad ...(13)$$

Thus we see that the probability that the neutron K.E. after the collision will lie in an interval dE centered around a kinetic energy E after collision is independent of this energy. In other words we can say that *all values of E are equally probable after a collision.* In eqn. (13) dW appears negative because dE is itself negative, representing, an energy loss. The probability per unit energy loss

$$P(E) = -\frac{dW}{dE} = \frac{1}{E_0}(1-\alpha). \qquad ...(14)$$

The average energy loss of the scattered neutrons after one collision

$$\Delta E = \frac{\int_{\alpha E_0}^{E_0}(E_0 - E)P(E)dE}{\int_{\alpha E_0}^{E_0} P(E)dE}.$$

Since the sum of the probabilities over all values of the energy is equal to unity, hence

$$\Delta E = \int_{\alpha E_0}^{E_0}(E_0 - E)P(E)dE = \frac{1}{2}E_0(1-\alpha).$$

$$\therefore \qquad \text{Average fractional energy loss} = \frac{1}{2}(1-\alpha). \qquad ...(15)$$

Thus we see that the same fraction of neutron energy is transferred to the moderator nucleus in each successive collision.

D. Transport Mean Free Path and Scattering Cross Section

It is clear that the scattering angles in the L-system are smaller than the scattering angle in the C.M.-system. It means in effect that the neutrons show a preferential forward scattering in the L-system, although the scattering is spherically symmetrical (isotopic) in the C.M. system.

The deviation from spherical symmetry is expressed in terms of the average value of cosines of all possible angles.

$$\cos\theta = \frac{\int_{\alpha E_0}^{E_0} \cos\theta\, P(E)dE}{\int_{\alpha E_0}^{E_0} P(E)dE} = \frac{\int_1^{-1} \cos\theta \left(-\frac{1}{2}\right) d(\cos\phi)}{\int_1^{-1} \left(-\frac{1}{2}\right) d(\cos\phi)} = \frac{2}{3A}.$$

Thus we see that for large A the average forward component becomes very small and scattering is almost isotropic. The forward scattering is predominant in the collisions with lighter nuclei. The predominant forward scattering in the L-system also affects the mean free path in the sense that the average distance travelled by the neutron before it is scattered through an angle 90° is greater than the corresponding average distance for isotropic scattering. The increased effective mean free path for nonisotropic scattering is called the *transport mean free path* λ_{ir} and is related to the *scattering mean free path* λ_s by the relation

$$\lambda_{tr} = \frac{\lambda_s}{1-\cos\theta} = \frac{\lambda_s}{1 - \frac{2}{3A}}. \quad ...(16)$$

$\therefore$ Transport cross section $\sigma_{tr} = \dfrac{1}{N_0 \lambda_{tr}} = \sigma_s \left(1 - \dfrac{2}{3A}\right)$ ...(17)

E. Average Logarithmic Energy Decrement

We shall now find the average value of decrease of the natural logarithm of neutron energy in a collision or *average logarithmic energy decrement,* usually denoted by ξ. According to the definition

$$\xi = \Delta \log E = \log E_0 - \log E = \log \frac{E_0}{E}. \quad ...(18)$$

This is done by dividing the probability of occurrence of a certain event by the probability of occurrence of all events of that type. If P(E) dE be the probability for a neutron to be scattered, in one collision, into an energy range lying between E and E + dE, then the probability that the logarithmic decrement $\log (E_0/E)$ be the one corresponding to a final energy E, when the initial energy was E_0, is given by the product $\log (E_0/E)\, P(E)\, dE$. If we assume the energy E to vary between its minimum (αE_0) and maximum (E_0) values, then

$$\xi = \frac{\int_{\alpha E_0}^{E_0} \log\left(\frac{E_0}{E}\right) P(E)\, dE}{\int_{\alpha E_0}^{E_0} P(E)dE}$$

$$= -\int_{\alpha E_0}^{E_0} \log\left(\frac{E}{E_0}\right) \frac{d\left(\frac{E}{E_0}\right)}{(1-\alpha)} = 1 + \frac{\alpha \log \alpha}{1-\alpha}.$$

In terms of the atomic mass number A, this becomes

$$\xi = 1 + \frac{(A-1)^2}{2A} \log \frac{A-1}{A+1}. \quad \text{...(19)}$$

For large values of A (> 10), it is easy to write

$$\xi = \frac{2}{\left(A + \frac{2}{3}\right)}. \quad \text{...(20)}$$

Thus we see that for very large A, ξ approaches zero. This is in line with our previous finding that heavy elements are poor moderators. We find a very good agreement if we compare the theoretical value with the experimental results.

When ξ is known, *the average number of collisions needed to bring about a given decrease in neutron energy*, can easily be calculated. If the neutrons start out with a average energy of E_1 and are slowed down to E_2 the total logarithmic energy loss is $\log\left(\frac{E_1}{E_2}\right)$. Then the average number of collisions is given by

$$n = \log \frac{\left(\frac{E_1}{E_2}\right)}{\xi}. \quad \text{...(21)}$$

F. Slowing-down Power and Moderating Ratio

For efficient slowing down we not only require a large value of ξ and a large probability of cross section for scattering and large N, the number of atoms per unit volume, but the *absorption cross section must be small* otherwise too many neutrons would be lost by absorption.

The three quantities ξ, σ_s and N jointly determine the slowing down ability of a moderator and their product is called the *slowing down power (sdp).*

$$\text{sdp} = \xi\sigma_s N = = \xi\Sigma_s = \frac{\xi}{\lambda_s}. \qquad ...(22)$$

Hence sdp can be interpreted as the average loss in logarithm of the energy per unit distance of travel in the moderator. Above eqn. indicates that a good moderator is one which has large values of Σ and of macroscopic scattering cross section Σ_s. The material chosen for the moderator should have a large sdp and should also have a low cross section for absorption. The ratio of sdp to macroscopic absorption cross section is the quantity which will indicate the better moderator, and is given the name of moderating ratio, represented by mr.

$$\text{mr} = \frac{\text{sdp}}{\Sigma_a} = \frac{\xi\Sigma_s}{\Sigma_\alpha}. \qquad ...(23)$$

When a moderator is not a one element substance, but consists of a combination of elements, the approximate value of ξ is given by

$$\xi = \frac{\xi_1\Sigma_s^1 + \xi_2\Sigma_s^2 + ...}{\Sigma_s^1 + \Sigma_s^2 + ...} \qquad ...(24)$$

$\therefore$ For a compound, $\text{sdp} = \xi_1\Sigma_s^1 + \xi_2\Sigma_s^2 + ...$ ...(25)

and $\text{mr} = \dfrac{\xi_1\Sigma_s^1 + \xi_2\Sigma_s^2 + ...}{\Sigma_a^1 + \Sigma_a^2 + ...}.$ ...(26)

G. Slowing-down Density

The rate at which the neutrons per unit volume of a moderator slow down past a particular energy E is called the slowing down density and is denoted by q(E). Owing to the discontinuous nature of the slowing down process and of the presence of absorption resonances for neutrons of discrete energies, the variation of slowing down density with energy is not expected to be simple. We can simplify the problem by treating the slowing down as a virtually continuous process, which is possible with sufficiently heavy moderator nuclei for which average energy loss per collision is very small.

If the rate of production of neutrons with an initial high energy E_0 is constant and equal to Q neutrons per unit volume per sec. If no neutrons are lost either by escaping (leakage) or by absorption before they have reached thermal energies, the slowing-down density q(E) will remain constant and will be equal to the source density Q. This is the case of an ideal moderator as we can only reduce the neutron escaping to negligible proportions by using a moderator of thick walls and the absorption is not zero, even if the moderator is infinite in extent.

The number of neutrons per unit volume whose energies lie between E and E + ΔE is equal to n(E) Δ E. If dt is the time needed by the neutrons to transit through the energy interval ΔE, the number of neutrons that are within this energy interval at a given time will be given by

$$n(E)\ \Delta E = -\ q(E)\ \Delta t. \qquad ...(27)$$

The negative sign is introduced because energy decreases with the increase of time. Here Δ(E) A E represents the neutron density crossing E per collision. The number of collisions this neutron group makes per second $= v/\lambda_s = v\Sigma_s$. Hence, the neutron density that crosses the energy E per second

$$q(E) = \{n(E)\ \Delta\ E\} \times v\Sigma_s. \qquad ...(28)$$

In the event of small energy changes in collision we can write

$$\frac{\Delta E}{E} = \Delta\left(\log\ E\right) = \xi \text{ or } \Delta E = E\xi.$$

$$\therefore \qquad q(E) = n(E)\ E\ \xi_v \Sigma_s, \qquad ...(29)$$

Thus in the absence of neutron absorption [q(E) = Q], the neutron flux per unit energy is given by

$$\phi(E) = n(E)\ v = \frac{Q}{E\xi\Sigma_s}. \qquad ...(30)$$

This expression shows that the slowing down neutron flux per unit energy is inversely proportional to the energy E.

The quantity $\phi(E)\ \Sigma_s$, is called *collision density* and is given by

$$F(E) = \phi(E)\Sigma_s = \frac{Q}{\xi E}. \qquad ...(31)$$

This relation shows that the scattering loss (no. of neutrons scattered out of the energy interval ΔE per sec. per unit volume) is equal to the neutron influx gain (the no. of neutrons scattered into the energy interval ΔE per sec. per unit volume). In the presence of absorption, q(E) will be related to Q as

$$q(E) = Q\,P(E), \qquad ...(32)$$

where P(E) is the probability or fraction of unabsorbed neutrons. The absorption loss $\Delta q = n(E)\,\Delta E\,v\Sigma_a = \phi(E)\,\Delta\,E\Sigma_a$. Hence in the steady state within the energy interval Δ E, Scattering loss + absorption loss = Influx

$$\phi(E)\,\Sigma_s\,\Delta E + \phi(E)\Sigma_a\,\Delta E = q(E)\,\frac{\Delta E}{E\xi}$$

or
$$\frac{\phi(E)\,\Sigma_a\,\Delta E}{\phi(E)\Sigma_a\,\Delta E + \phi(E)\Sigma_s\Delta E} = \frac{\Delta q E\xi}{q(E)\Delta E}$$

or
$$\left[\frac{\Sigma_a}{(\Sigma_a + \Sigma_s)}\right]\left[\frac{\Delta E}{E\xi}\right] = \frac{\Delta q}{q(E)}. \qquad ...(33)$$

H. Slowing Down Time

The time required for the neutrons to slow down from an initial energy E_0 to a final energy E is known as the slowing down time, can be obtained by eqns. (27) and (29).

$$\therefore \qquad \Delta t = -\frac{\Delta E}{E\xi\;v\Sigma_s}. \qquad ...(34)$$

If we assume that ξ and Σ_s remain constant over the slowing down range of energy, the above expression can be integrated as

$$\text{Slowing down time } T = \int_0^T dt = -\frac{1}{\xi\Sigma_s}\int_{E_0}^{E_f}\frac{dE}{Ev}$$

$$= -\frac{\left(\frac{1}{2}m\right)^{1/2}}{\xi\Sigma_s}\int_{E_0}^{E_f} E^{-3/2}dE$$

$$= \frac{(2m)^{1/3}}{\xi\Sigma_s}\left[\frac{1}{E_f^{1/2}} - \frac{1}{E_0^{1/2}}\right] \qquad ...(35)$$

With the help of this relation we can also calculate the average slowing-down time of neutrons for various moderators.

ENERGY DISTRIBUTION OF THERMAL NEUTRONS

The slowing-down process is due to an elastic scattering process as emphasised earlier. The moderating material, used for this slowing down process, should have a large scattering cross-section and small absorption cross-section for neutrons. As a result of the slowing-down process, neutrons reach the state in which their energies are in equilibrium with those of the atoms or molecules of the moderator in which they are moving. At this time their energy distribution will be approximately Maxwellian, corresponding to the temperature of the surrounding medium. In thermal equilibrium the velocity distribution of neutrons will be given by the expression

$$dn = n(v)\, dv = 4\pi n \left(\frac{m}{2\pi kT}\right)^{3/2} v^2 e^{-mv^2/2kT} dv, \qquad ...(36)$$

where n is the total number of neutrons per unit volume, m is the mass of the neutron, T the absolute temperature and k the Boltzmann constant. The number of neutrons whose velocities lie between v and v + dv is given by dn = n(v) dv. The velocity distribution function (Fig. 4.12) has a maximum for a value of the velocity v_p, which is called the most *probable velocity*. It can be determined by the usual procedure of differentiating the right hand side of eqn. (36) with respect to v and setting equal to zero. In this way we have $v_p = (2kT/m)^{1/2}$. The energy corresponding to this velocity is denoted by E_p and is given by

$$E_v = \frac{1}{2} m v_p^2 = kT. \qquad ...(37)$$

The energy distribution of the neutrons is given by

$$dn = n(E)\, dE = 2\pi n\, (\pi kT)^{-3/2}\, E^{1/2}\, e^{(-E/KT)}\, dE, \qquad ...(38)$$

and is shown in Fig. 4.13. This distribution leads to a most probable energy E_0, which can be found by differentiating eqn. (38) and equating this to zero. The value thus obtained is 1/2kT, which is one half of the energy E_p, the energy corresponding to the most probable velocity. The velocity corresponding to this energy is known as average velocity and is given as

$$\bar{v} = \left(\frac{2}{\pi^{1/2}}\right) v_p = 1.128\, v_p.$$

The average energy E is given by kinetic theory as

$$\bar{E} = \frac{3}{2} kT = \frac{3}{2} E_0. \qquad ...(39)$$

This average energy does not correspond to the average velocity but to the root mean square velocity v_{rms}, given by

$$\overline{E} = \frac{1}{2}mv^2 = \frac{1}{2}mv_{rms}^2$$

or $$v_{rms} = \left(\frac{3kT}{m}\right)^{1/2} = \left(\frac{3}{2}\right)^{1/2} v_p. \qquad ...(40)$$

The velocity distributions of thermal neutrons from different sources have been measured experimentally. For various experimental reasons, highly accurate measurements are hard to make, but the measured velocities do follow Maxwell distribution within the experimental accuracy of about 10%. This discrepancy corresponds to a temperature which is slightly greater than that of the moderator material. The steady influx of high energy neutrons and at the same time a steady absorption of the low energy neutrons raises the high energy end of the Maxwell distribution and depresses the low energy portion of the distribution.

NEUTRON VELOCITY SELECTOR

Even accelerator sources do not provide monoenergetic neutrons for accurate cross-section analysis. The problem has been solved with the aid of devices for selecting monoenergetic neutrons from heterogeneous neutron sources. These are known as velocity selectors. We now consider some of these devices.

1. Crystal Spectrometer

For a thermal and epithermal neutrons ($E < 10$ eV) a moroenergetic neutron beam can be selected by diffraction from a single crystal. The design is based on the wave properties cf neutrons. It can be seen from relation $A = h/mv = h/\sqrt{(2mE)}$ that neutrons with energies in the range 0.01 eV to about 10 eV have wavelengths of the order of magnitude as the wavelengths of X-rays and of the distance between crystal lattice planes. When a beam of slow neutrons of wavelength λ is reflected from the surface of a crystal, the diffraction maxima are observed at angles given by the Bragg relation

$$2d \sin \theta = n\lambda, \qquad ...(1)$$

where n is an integer which represents the order of the reflection, d is the spacing of the lattice planes and θ is the glancing angle of the incident neutrons for with order reflection. If the de Broglie relation is used, the velocity of the neutrons at the nth maxima is given by

$$v = \frac{h}{m\lambda} = \frac{nh}{2md} \sin\theta \qquad ...(2)$$

$$\therefore \qquad E = \frac{n^2h^2}{8md^2} \sin^2\theta. \qquad ...(3)$$

A systematic diagram of a crystal spectrometer. A well collimated beam of thermal neutrons from an intense source, *e.g.*, a nuclear fission reactor, is diffracted by a prominent set of lattice planes of a large well shielded single crystal. The diffracted neutrons enter a second long collimator before entering a well shielded BF_3 cylindrical proportional counter.

Neutrons of various speeds are present in the beam but for a particular value of θ, the great majority of the neutrons reaching the detector will have a velocity given by equation (2). If the glancing angle θ is changed, neutrons with a different velocity will now reach the detector. As the glancing angle diminishes with increasing neutron velocity, the angle ultimately becomes too small to be measured. This sets an upper limit to the speed of the neutron which can be studied in the crystal spectrometer.

The energy resolving power dE/E decreases with increasing neutron energy. The differentiation of equation (2) gives

$$dE = \frac{n^2h^2}{8md^2} \frac{2}{\sin^3\theta} \cos\theta \, d\theta$$

$$\therefore \qquad \frac{dE}{E} = -2\cot\theta \, d\theta.$$

For small θ, dE/E = –2 dθ/θ. ...(4)

For 1 eV neutrons reflected from the crystal (d = 2.32 Å) in the first order, 0=3.5°. If the sharpness of collimation determines dθ = 0.1°, the resolution dE/E ~ 6%. The main drawback of the crystal diffraction method is the occurrence of second order reflection of neutrons of wavelength λ/2 at the same angle as first order neutrons of wavelength λ.

2. Mechanical Velocity Selectors

The design of this type of chopper is based on the fact that cadmium is a strong absorber of neutrons with energies less than 0.3 eV, while other metals such as aluminium are weak absorbers in this energy region. A cylinder, made up of alternate layers of cadmium and aluminium running parallel to the axis, is rotated before a strong source of heterogeneous neutrons. This device acts as a shutter and passes neutrons through it only when the laminations are parallel to the neutron direction. Two bursts of neutrons are released at each revolution. Neutrons then arrive at a detector after travelling a distance/and their time of flight t depends on their velocity $v = l/t$. The detector must be sensitive for a short period at a definite interval after the neutrons have been transmitted. The timing is achieved by means of a double sided mirror fixed to the axis of the rotating cylinder. This reflects a beam of light from a fixed lamp into a photocell at an instant between successive bursts given by the angular setting of the cell. The time t is determined by the speed of rotation of the cylinder and also by the relative positions of source and cell. In the slow neutron chopper of Fermi, J. Marshall and L. Marshall, the BF_3 neutron counter was placed 1.46 metres beyond the rotating cylinder. The neutrons detected are those with a speed of l/t. The cross-section can be obtained by introducing the absorbing material at A and can be determined as a function of energy.

Above about 0.3 eV the Cd sheets become too transparent and a more satisfactory chopper uses materials with a large scattering cross-section, *e.g.,* nickel steel or nickel coated cadmium. Fast choppers are more complicated and difficult to operate. In one form of fast neutron chopper two steel cylinders, about 16" long and 4" diameter are mounted horizontally end to end in front of the neutron source. Each cylinder has a number of narrow slits on the outside, running parallel to the axis. One cylinder is stationary and the other is rotated at high speed. This type of fast chopper can be used with neutrons having energies up to about 5 keV.

3. Time of flight Velocity Selector

It is necessary to measure cross-sections at neutron energies greater than those covered by the slow chopper and crystal spectrometer. The time of flight method for measuring the neutron velocity can be used with a cyclotron or linear accelerator which produces a neutron beam. Fast neutrons are slowed down in paraffin block and then allowed to pass through a thin slab of material. The time required for the neutrons to reach a detector about 10 metres away from the source is electronically measured. The timing is usually done by distributing the counter pulses to a series of channels according to time. In this way neutrons of certain velocities are selected from a non-homogeneous beam. Transmission curve obtained with an instrument used at Columbia University is shown in Fig. 4.19. The dips in the transmission curve correspond to the resonances. This instrument can be used for neutron energies from 0.003 to 5000 eV.

One of the best methods of producing short intense neutron bursts in the pulsed electron accelerator. The latest machine of this type installed at Harwell (1960) accelerates pulses of electrons (~ 0.25 μ sec) to an energy of 30 MeV. They strike a high atomic number target which serves the dual purpose of producing energetic bremestrahlung, which in turn liberates neutrons through the (γ, n) reaction.

DISCOVERY OF NEUTON

Rutherford in his Bakerian Lecture to the Royal Society in 1920 said, "*Under some conditions, ..if may be possible for an electron to combine much more closely with the hydrogen nucleus (i.e., a* ***proton),*** *forming a kind of neutral doublet. Such an atom would have* ***novel*** *properties.*

Its external field would be practically zero...and consequently it should be able to move freely through matter. Its presence would probably he difficult to detect." In 1930, Bothe and Becker bombarded lithium, beryllium and boron with α-particles from polonium and found a very penetrating but nonionizing radiation, they assumed that the radiation was of the γ-ray type because of its tremendous penetration.

While repeating these experiments in 1932, Dr. H. Joilot and his wife, Dr. Irene Curie Joliot, daughter of Madam Curie found that which a sheet of hydrogen containing material, particularly paraffin, was interposed in the path of these radiations, protons were ejected with a considerable velocity. From the ranges of these recoil protons, the maximum proton energy E proved to be about 5.3 MeV.

Assuming that the protons were produced as the result of elastic collisions with the *γ-ray photons*, calculations showed that each photon must have possessed an amount of energy of about 52 MeV. These results were entirely inconsistent with the results from the experiments on the absorption of these rays in lead (about 7 MeV). For an α-particle of 5 MeV energy :

$$_4Be^2 + {_2He^4} + E_a \rightarrow (_6C^{13}) \rightarrow {_6C^{12}} + h\nu$$

$$9.01503 + 4.00388 + .00536 = 13.00751 + h\nu$$

or $h\nu = 0.01676$ amu, $= 15.5$ MeV.

This is much shorter than the required energy in order to explain the proton scattering. At Cambridge, Webster observed that the energy of the photons emitted by the beryllium varied according to their direction of emission, heaving least value if emitted in the direction opposite to that of the incident α-particles. The situation was resolved by Chadwick, in England in 1932, after performing a series of measurements of the energies of recoil of protons ejected from thin targets by the penetrating "Be-radiation" with a pulse ionizing chamber and amplifier.

He said, *"These results are very difficult to explain on the assumption that the radiation from beryllium is a quantum radiation, if energy and momentum are to be conserved in the collisions. The difficulties disappear, however, if it be assumed the radiation consists of particles of mass* 1 *and charge 0, or neutrons."* Chadwick's theory made sense of Webster's puzzling discovery that the neutron energy varied with the direction of emission. These neutrons were formed as a result of the highly exoergic nuclear reaction

$$_4Be^9 + {}_2He^4 \rightarrow ({}_6C^{13}) \rightarrow {}_6C^{12} + {}_0n^1 + 5.65 \text{ MeV}.$$

Thus we see that *comparatively late date of the discovery of the neutron is partly due to the similarity in the properties of the neutron and high energy photons and the confusing effect that γ-rays as well as neutrons are associated usually with neutron sources.*

NEUTRON SOURCES

1. Photo-neutron Sources

When γ-rays of sufficient energy (greater than the neutron binding energy) are incident upon light elements a probability exists that neutrons will be emitted by the energized nuclei. The majority of nuclei have a neutron binding energy of 5 MeV or more, except $_4Be^9$ (1.67 MeV) and $_1H^2$ (2.23 MeV). g-rays from artificially produced radio-nuclei can serve in photo disintegration sources.

A standard source consisting of a *MsTh* or Na^{24} gamma source is placed at the centre of a spherical vessel filled with heavy water as shown in Fig. 4.1. Yield, the number of neutrons per sec due to 1 gm. of Be or D_2O at a distance of 1 cm. from 1 curie, of various gamma sources is given in the table 4.1.

Table 4.1 : Radioactive (γ, n) sources.

γ-Source	Target	Half Life	E_γ (MeV)	E_n (MeV)	Yield
Na^{24}	Be	14.8 hr.	2.76	0.83	13×10^4
Na^{24}	D_2O	14.8 hr.	2.76	0.22	27×10^4
Y^{88}	Be	87 d.	1.9, 2.8	0.16	10×10^4
Y^{88}	D_2O	87 d.	2.8	0.31	0.3×10^4
Sb^{12}	Be	60 d.	1.7	0.025	19×10^{24}

The advantages of this type of source are :

(a) The reproducible neutron strength,

(b) The choice of E_n by choosing a suitable value of E_γ.

(c) The wide variety of artificial γ-sources available from atomic reactors,

(d) The neutrons produced are more mono-energetic than those obtained from (Be – α) sources.

Wattenberg gave the equation for the energy E_γ of the emitted neutrons as

$$E_n = \frac{A-1}{A}\left[E_\gamma - Q - \frac{E_\gamma^2}{1862(A-1)}\right] + E_\gamma \cos\theta \left[\frac{2(A-1)\left(E_\gamma - Q\right)}{931\,A^3}\right]^{1/2}$$

where A is the mass of the target nucleus, E_γ the energy of γ-ray in MeV, the threshold energy in MeV for the (γ, n) reaction in the target nucleus and θ is the angle between the path of the γ-ray and the direction of the emitted neutron. As Na^{24}-source is of short half life and hence Sb^{124} is very convenient for intermediate energy neutrons with a relatively long half life.

2. Radioactive Sources

The majority of the neutrons is produced by the (a, n) reaction given above. If the radium is in secular equilibrium with its decay products, extra deuterons are released according to the reaction

$$_4Be^9 + h\nu \rightarrow {_4Be^8} + {_9n^1}.$$

The product nucleus of this reaction Be^8, is unstable and thus decays into two a-particles within 10^{-15} sec. This neutron source is isotropic and produces neutrons over a wide energy range (0 → 13 MeV). The neutrons are not mono-energetic due to following reasons:

(a) The α-particles do not all have the same initial energy.

(b) Many α-particles lose part of their initial energy by ionization before nuclear capture,

(c) The neutron energy varies with the direction of emission having least value if emitted in the direction opposite to that of the incident α-particles.

(d) ${}_6C^{12}$ can be left in an excited state.

We can have slow neutrons after putting the Ra – α – Be source in water, paraffin, or some other hydrogen-containing material. For a 2-MeV neutron only 25 elastic collisions with protons are needed. Radon is some times used instead of radium, as it is a gas and can be made into a more compact source. But it has the disadvantage of decaying rapidly because of the snort half life of 3.8 days. Polonium is used instead of radium when a neutron source with relatively few g-rays is desired. This has several disadvantages :

(i) the neutron output is much lower (3×10^6 neutrons/sec per curie of Po while it is $10 - 18 \times 10^6$ neutrons/sec per curie of Ra for radium),

(ii) special chemical facilities are needed to prepare Po,

(iii) the half life is short (140 days) — compared with that of Ra – Be source (1600 years).

3. Reactor Sources

The most powerful source of neutrons available in these days is a nuclear reactor or chain reacting pile on which uranium undergoes fission. The fission of each heavy nucleus by slow neutron capture is accompanied by the release on the average of about 2.5 neutrons. The neutrons produced from the fission on of uranium atoms are fast and are slowed down in the moderator. The disadvantages of reactor sources are the wide spread in neutron energy and the impossibility of directly pulsing the neutron source.

4. X-rays Sources

Through the (γ, n) reaction, high energy X-ray machines can produce the high energy neutrons. A flux of high energy photons is produced when the electrons accelerated to high energies in linear accelerators, betatrons, synchrotrons etc. impinge on target elements of high atomic number. A large number of neutrons within the target material are released by the high flux of photons. The disadvantage of this type of source is the wide spread in neutron energy.

5. Ultrafast Neutrons

As a deuteron behaves as a relatively loosely bound system consisting of a neutron and a proton which are frequently outside the range of their mutual forces, hence, deuterons accelerated to about 200 MeV in a synchro-cyclotron can readily be "stripped" into protons and neutrons by directing the deuteron beam to pass through a target nucleus. Recently neutrons of energy > 300 MeV have been produced by stripping 650 MeV deuterons accelerated in the Birmingham synchrotron. Almost any element may be used as the target material for stripping out the yield varies with its nature.

6. Neutrons from Accelerated Particle Reactions

Neutrons are produced by (d, n) and (p, n) reactions in particle accelerators. These sources give mono-energetic neutrons in the range from a few keV to 20 MeV. In deuteron bombardment the energy of the released neutron is progressively larger for most of the target elements, which is due to the low binding energy of the deuteron. The C^{12} (d, n) N^{13} reaction is an exception to the rule with a Q value of – 0.28 MeV. Those neutrons which are emitted in the direction of the incident beam of deuterons will have a maximum energy, equal to the sum of the incident and the reaction energies approximately. Some (d, n) reactions are:

$$_1H^1 + d \rightarrow {_2He^3} + {_0n^1} + 3.265 \text{ MeV}$$

$$_1H^3 + d \rightarrow {_2He^4} + {_0n^1} + 17.6 \text{ MeV}$$

$$_3Li^7 + d \rightarrow {_4Be^8} + {_0n^1} + 15.0 \text{ MeV}$$

$$_4Be^9 + d \rightarrow {_5B^{10}} + {_0n^1} + 3.79 \text{ MeV.}$$

Several light and moderately heavy elements undergo (p, n) reactions. These endoergic reactions are the convenient sources of neutrons of moderate energy. Some (p, n) reactions are :

$$_1H^3 + p \rightarrow {_2He^3} + {_0n^1} - 0.764 \text{ MeV}$$

$$_3Li^7 + p \rightarrow {_4Be^7} + {_0n^1} - 1.646 \text{ MeV.}$$

The Li^7 (p, n) Be^7 reaction suggests that accelerated charged particles can be used to produce nearly monoenergetic neutrons.

NEUTRON COLLIMATORS

Neutrons produced by the above sources possess very high energy range. Most of the neutrons are called fast neutrons as they have enormous velocities. Since the ordinary methods used for charged particles and γ-rays can not be used for the collimating a beam of fast neutrons. Dunning (1938) has devised a collimator known as the *neutron howitzer*. He used the property of slowing down of neutron in hydrogeneous materials like paraffin or water. The radium beryllium source S is placed at the bottom of a barrel inside a paraffin block. A cylindrical slit is cut in this block which is lined from all sides by cadmium sheet. Neutrons are emitted in all directions. The neutrons energy is negligibly small in all directions except through the barrel. The thermal neutrons resulting from the slowing down of the fast neutrons in the paraffin are effectively absorbed by the cadmium coating over the outer surface a paraffin and walls of the barrel. The cadmium layer is covered by a layer of lead to g-rays arising from neutron capture.

BASIC PROPERTIES OF THE NEUTRON

1. Neutron Charge

The neutron is usually assumed to have no net charge. Dee investigated the ionization proceed in air in a cloud chamber irradiated by fast neutrons and found one ion pair per 3 metre of the path and thus concluded that the charge of the one neutron is less than 1/700 of the proton charge. Rabi and his coworkers have found the neutron charges less than 10^{-12} electron charge. Fermi and Marshall set an upper limit of ~ 10^{-18} electron charge for the charge on the neutron.

2. Neutron Mass

Chadwick gave a rough estimation of the mass of a neutron from the information obtained in collisions of neutrons with protons on the one hand and with nitrogen nuclei on the other. From the equations for the conservation of energy and for the conservation of momentum for a head on collision, and from the lengths of accompanying tracks in the cloud chamber, Chadwick found that neutron mass M_n = 1.15 amu, roughly the same as that of the proton.

A more precise value for the mass of the neutron was obtained by Chadwick from the study of the nuclear reaction

$$_5B^{11} + {}_2He^4 \rightarrow {}_7N^{14} + {}_0n^1 + Q.$$

The Q-value was found from the kinetic energies of B^{11}, N^{14}, $_0n^1$ and α-particles. Substituting all the known values in terms of atomic mass units in the above equation Chadwick obtained 1.0067 mu. for the mass of the neutron.

J. Chadwick and M. Goldhaber in 1934 investigated the photo-nuclear reaction ($_1H^2 + h\nu \rightarrow {}_1H^1 + {}_0n^1$) using gamma rays of ThC", with $h\nu = 2.62 \times 10^6$ eV. Since the mass of the deuteron and of the proton were known, the only required measurement was the kinetic energy of the proton and of the neutron.

The proton energy was measured by an ionization method and found to be almost 1.05 MeV. If it is assumed that the neutron and proton mass are nearly equal than the disintegration energy is nearly equal to twice the recoil energy of the proton. Using the best mass-spectrographic atomic masses of hydrogen and deuterium and substituting the mass equivalents for the γ-ray energy and disintegration energy (proton energy + neutron energy), they calculated

$$\text{Mass of neutron} = 1.0087 \pm 0.0003 \text{ amu.}$$

The most precise determination of the mass of the neutron was later made by Bell and Elliott from the reaction

$$_0n^1 + {}_1H^1 \rightarrow {}_1H^2 + h\nu.$$

They carefully measured the energy of the γ–ραψσ and obtained a value of 2.230 ± 0.007 MeV. Therefore, the mass of the neutron from above equation becomes 1.008672 mu.

3. Neutron Statistics and Spin

Since the deuteron is a loose combination of proton and neutron and is supposed to have a spin 1. From the study of band spectra of deuterium and the proton spin (1/2) the neutron spin comes out to 1/2. Neutrons have been found to obey Fermi statistics.

The experimental fact that the magnetic moment of the deuteron is nearly equal to the algebraic sum of the magnetic moments of the neutron, and proton strongly supports this value. This value of 1/2 for the spin of the neutron has been confirmed by reflection of neutrons from magnetised mirrors by Hughes and Burgy.

4. Decay of the Neutron

The first accurate determination of the mass of neutron by Chandwick and Goldhaber showed that the neutron mass was greater than that of a proton. It was also evident that the neutron outside the nucleus is unstable and could be expected to decay with the emission of a beta-particle plus a neutrino, leaving a proton. It was also assumed that a similar process accounted for the β-emission of β-unstable radioactive nuclei.

$$n \rightarrow p + e^- + \overline{v} + 782 \text{ keV}.$$

Using the neutrons from the Oak Ridge pile, A.H. Snell and his coworkers (1948) were able to detect proton coming out from the beam of neutrons. The presence of electrons were also shown by them after some time. J.M. Robson (1951), using the much more intense source of neutrons available from the Chalk River Pile, not only observed the decay of the neutrons but also succeeded in measuring the energy spectrum of the decay electrons. The plan of the experimental arrangement is shown in Fig. 4.3.

A collimated beam of pile neutrons from a nuclear reactor was allowed to pass into a vacuum chamber. Some of the neutrons decayed in their fight through the vacuum chamber. The decay protons, with a small fraction of the disintegration energy, were deflected by the deflector (high voltage electrode), maintained at a potential of + 13 kV with respect to the earthed vacuum chamber, into a proton spectrometer and detected with an electron multiplier. The β-rays emitted during this disintegration were deflected to the left into a β-rays spectrometer and were detected with a scintillation counter using anthracene crystals as the phosphor.

These two detectors were connected to a coincidence circuit which would register only when pulses from the two detectors arrived there simultaneously. Owing to its greater mass the proton would move more slowly through its spectrometer system and would be delayed with respect to the electrons. Calculations gave this delay as about 0.9 micro second.

The rate of proton-beta particle coincidences is plotted against the energy of the β-particles, measured in the β-spectrometer, as shown in Fig. 4.4. Extrapolation of the curve gives a value for the maximum energy of the spectrum as 782 ± 13 keV, which is the close approximation of the mass difference between the neutron and the proton.

5. Half Life of Neutron

The estimate of the half life of the neutron was obtained from a determination of the density of the neutron beam and the number of neutrons decaying per unit time per unit volume. The neutron density was obtained by exposing manganese foils to the neutron beam and measuring their activities against the activities of the calibrated standard foils of identical thickness. At the centre of the beam the value so obtained was 1.6 × 1010 neutrons/m^3. The number of neutrons decaying per unit time per unit volume was estimated from the number of protons per unit time striking the first electrode of the electron multiplier and the volume of the beam where the protons originate. The value so obtained was 6.3 × 108 neutrons/min m^3 of the neutrons beam. Therefore, the half life T becomes

$$T = 1.16 \times 10^{10} \times 0.693/6.3 \times 10^{8} = 12.8 \text{ min.}$$

Robson's observations gave a value for the half life T as 12.8 ± 2.5 min, Recently, a more accurate determination of T by a Russian group by a similar method gave the value T = 11.3 ± 0.3 min. An alternative method was used by D. Angelo at the Argonne National Laboratory in the same year. His result is

$$T = 12.7 \pm 1.9 \text{ min.}$$

6. Wave Properties

For non-relativistic velocities ($v << c$), the wavelength associated with neutrons is given by

$$\lambda = \frac{h}{p} = \frac{h}{(2ME)^{1/2}} = \frac{2.86\times10^{-11}}{E^{1/2}} \text{ metre,}$$

where E is in eV. Thus for thermal neutrons A = 1.82 A° which is comparable with atomic dimensions. For E = 1 MeV, $\lambda = 2.86 \times 10^{-14}$ metre approaching nuclear dimensions. The relativistic correction is important above energy of 100 MeV. The wave properties are of primary importance in determining the nature of the interaction between neutrons and nuclei.

7. Magnetic Moment of the Neutron

It is sometimes said that the neutron should not have a magnetic moment because it has zero charge. Actually magnetic moments depend upon currents rather than charges. For example the hydrogen atom has no net charge but does have a magnetic moment. Thus neutron can be supposed to have a magnetic moment if it contains equal +ve and –ve charges which are in motion. According to Yukawa's meson field theory neutron transforms into a proton and a negative π-meson for a fraction of time. Since this meson is lighter than the proton, its motion will be faster and the resultant magnetic moment is that of revolving a negative charge. Hence it is negative. The proton and π-meson also possess spin and, therefore, have intrinsic magnetic moments, that of the meson being large due its small mass. This situation also contributes to a negative moment of the neutron.

Nuclear resonance method in the original form cannot be applied here because it is not possible to have neutrons in large densities just like protons. The value of magnetic moment of neutron was measured with considerable accuracy by Alvarez and Bloch in 1940. They used essentially the magnetic resonance beam method employing solid magnetized steel plates as polarizer and analyzer in place of the magnets.

In an experiment used by Bloch and others (Fig. 4.5), neutrons produced in a beryllium target by cyclotron accelerated deuterons are slowed down in a paraffin block, then emerged through a hole in a 9-in-shielding wall and moved as a collimated beam through a sequence of three magnets (all with magnetic fields directed upward). The neutrons in the beam, after passing through the highly saturated iron piece present in the gap of the first magnet, are partly polarized (down) because more spin up than spin down neutrons have been scattered out.

In the absence of second magnet, some of the neutrons in the beam when passing through the iron piece present in the gap of the third magnet, are scattered out of it. The amount of reduction of the beam intensity depends on the magnitude of the polarization previously attained. The relative loss in the third magnet is less than in the first since more than half of the neutrons in the beam hitting the scatterer have spin down. An R.F. field with field direction perpendicular to the main field is produced in the gap of the second magnet. When this field equals the Larmor frequency of the neutrons in this magnet such that the neutron spins can be flipped, the effect is to destroy fully or partly the beam polarization and the result is a reduction in intensity observed at the counter. For a constant field in the precession magnet, the frequency of the R.F. field was varied and the quantity

$$E_T = \frac{(I_{on} - I_{off})}{(I_{off} - I_{bg})} \qquad ...(1)$$

was measured. Here I_{on} and I_{off} are the counting rates with the R.F. field on and off respectively and I_{bg} is the background counting rate.

Instead of attempting to measure directly the field induction B in the precession magnet and thus determining μ, Bloch and others measured the ratio $|\mu_n|\mu_n$ (= ν_n/ν_p). Here ν_n and ν_p are the larmor frequencies in the same field for neutron and proton respectively. They made use of the fact that the ratio $\nu_p\ \nu_n \sim 2/3$. The signal used in the proton resonance apparatus was the sixth harmonic of a master oscillator of frequency $\nu_p' = \nu_p/6$. Similarly the frequency of another mass oscillator was $\nu_n' = \nu_n/4$. We thus have

$$\frac{|\mu_0|}{m_p} = \frac{\nu_n}{\nu_p} = \frac{4\nu_n'}{6\nu_p'} = \frac{2}{3}\left(1 + \frac{\Delta\nu_n'}{\nu_p'}\right) \qquad ...(2)$$

Here $\Delta\nu_n' \left(= \nu_n' - \nu_p'\right)$ is the difference between the two master oscillators. It was measured very accurately. A typical neutron resonance observed by Bloch and others is represented by Fig. 4.6. For this curve we have maximum R.F., field BRF = 11.5 gauss. $\nu_p' = 7.674$ Mc/sec and $\Delta\nu_n' = 210.5$ kc/sec. at resonance. We thus have

$$\frac{|\mu_n|}{\mu_p} = \frac{2}{3}\left(1 + \frac{0.2105}{7.674}\right) = 0.6850. \qquad ...(3)$$

Using the results for μ_p, obtained in chapter 1, we have

$$|\mu_n| \simeq 1.91303 \text{ nuclear magneton.}$$

A better method, achieving a much higher degree of polarization is to reflect the beam at a very small glancing angle from a magnetized mirror. Ramsay and others used this technique and were able to measure the neutron magnetic moment accurately. Neutron beams from cyclotron induced reaction (d, n) or high flux reactors are convenient sources. The neutrons are reduced to thermal velocities by a hydrogeneous moderator before they are collimated by cadmium sheets. This collimated beam of thermal neutrons (700,000 counts/rein.) is incident at a glancing angle of 17' on a magnetic mirror used as a polarizer. An alloy of 93% Co and 7% Fe is a suitable material for the mirror. The reflected beam has a polarization as high as 90%. The beam of polarized neutrons passes through a 5 ft long uniform magnetic field H_0 (~ 8500 gauss). The beam is then reflected at a small angle from a second magnetic mirror used as a analyzer (magnetized in the same direction as first mirror). The intensity of the doubly reflected beam (~ 7500 counts/min) is reflected by BF_3 proportional counter.

Ramsay has previously shown that the method of using two separated current loops produces narrower resonance lines. The neutron spins in the long magnetic field are flipped by two radio-frequency current loops mounted 42 inches apart. Resonance flipping will occur when the angular frequency of the radio-frequency supply coincides with the Larmor angular frequency of the neutron spins. The spin flipping reverses the polarization of the beam. Thus the resonance is revealed by a sharp decrease in the number of neutrons detected by the counter. Similar to the molecular beam resonance experiments the resonance condition for spin half particles is given by

$$2m_n B_0 = \hbar\omega \text{ or } \mu_n = \frac{\hbar\omega}{2B_0}. \qquad ...(4)$$

The magnetic field B_0 is conveniently measured using a proton resonance probe. The value of the neutron magnetic moment obtained by Ramsay, Corngold and Cohen is

$$\mu_n = -1.913148 \pm 0.000066 \text{ nm.}$$

The – ve sign indicates that the magnetic moment of the neutron is similar to that generated by rotating – ve charge with its spin parallel to that of the neutron.

Since the deuteron is believed to consist of a proton and a neutron with parallel spins, the magnetic moment should be equal to the sum of the separate nucleon moments. But this sum differs from deuteron moment by 0.022 nuclear magneton. This discrepancy may be due to following reasons:

(a) The deuteron spends part of its time in a state of higher orbital momentum, in which its magnetic moment would be definitely less than the ground state.

(b) There would be a difference in the magnetic moments of the nucleons in the free and combined conditions.

CLASSIFICATION OF NEUTRONS AS TO ENERGY

The neutron energy can be determined by measuring the ranges projected by them in an ionisation chamber. For more detailed examination of the results of the interactions of neutrons with matter it is also desirable to classify them according to their kinetic energy.

1. Ultra Fast Neutrons

Neutrons with energies beyond 50 MeV are called *ultra high energy neutrons*. They are produced in the target by p-n interactions induced in nuclei by high energy protons. The cosmic radiation is also a source of neutrons with energies well above those which are likely to be producer by accelerations.

2. Slow Neutrons

Neutrons with energies from zero to about 1000 eV are usually included in this category. In this range sub-classifications occur. The most important are given below.

(a) Epithermal Neutrons

Consider an arrangement in which a fast neutron source is placed inside a moderator which slows the neutrons down until they are in equilibrium with the molecules of moderator. Before establishing the thermal equilibrium, the distribution of velocities will contain velocities which exceed any permitted by a Maxwell distribution for the temperature of the moderator. Such a distribution is called *epithermal* and the neutrons in it are called *epithermal neutrons*.

(b) Cold Neutrons

Neutrons having their average energy less than thermal neutrons are called cold neutrons. These neutrons are produced by a device depending on the coherent scattering. From the Bragg law ($n\lambda = 2d \sin \theta$) we have no reflection if the neutron wavelength λ exceeds 2d, where d is the grating spacing of the crystal.

Thus the neutrons with the lower energies are transmitted. Since 2d for graphite is 6.7 A° the computed maximum energy of the transmitted neutrons is 0.002 eV, which is well below the average energy of thermal neutrons.

(c) Resonance Neutrons

Since the various nuclei exhibit strong absorption of neutrons at fairly well defined energies in the range of energies between 1 to 100 eV. These absorptions are known as resonance absorptions and the neutrons having the corresponding energies are known as resonance neutrons.

(d) Thermal Neutrons

When fast neutrons have been slowed down until the average energy of the neutrons is equal to the average thermal energy of the atoms around them, the neutrons are called *thermal neutrons*. At each collision with these atoms, the neutron may gain or loss energy. They are in thermal equilibrium with their surroundings and they may be assumed to have a velocity distribution analogous to Maxwell distribution for the molecules of a gas.

$$dn(v) = Av^2 e^{-Mv^2/2kT_{dv}}, \qquad ...(1)$$

where v is the neutron velocity, M its mass and T the absolute temperature. The maximum number of neutrons will have the energy kT, which is about 0.025 eV at 20°C.

3. Very Fast Neutrons

This energy interval (10-50 MeV) is distinguished from the proceeding by the appearance of nuclear reactions involving the emission of more than one product, such as the (n, 2n) reaction.

4. Intermediate Neutrons

Intermediate neutrons in the energy region between 1000 eV and 0.5 MeV are obtained by the deceleration of fast neutrons. We have less informations about intermediate neutrons than about slow neutrons because of the difficulty of finding efficient detectors. In this energy range elastic scattering process is dominant. Recently a number of techniques have been developed for the study of this energy range.

5. Fast Neutrons

Neutrons with energies having range between 0.5 to 10 MeV are called fast neutrons. This energy region is characterised by the appearance of many nuclear reactions which are energetically impossible at lower neutron energies of which the most important is inelastic scattering.

NEUTRON DIFFUSION

Diffusion is the process that takes place when particles of one type move between particles of another type into which they are penetrating. In the previous article we focused our attention on the energy loss suffered by the neutrons as a consequence of collisions with the nuclei of the moderator. It is often necessary to know the spatial distribution of the neutrons, *i.e.,* the dependence of the neutron density on position. The problem can be treated with the aid of the theory of diffusion. This is not new in physics but is similar to the problem of the diffusion of a gas through another gas or to that of the diffusion of electrons though a gas. For the sake of simplicity we make several assumptions (i). The neutron flow does not change with (ii) We assume that neutron absorption is negligibly small (iii) In lab. system, the probability of a particular scattering angle is a constant. (iv) The neutrons are monoenergetic and the scattering is entirely due to elastic collisions. The result of calculations based on these assumptions are in fairly good agreement with experiment.

(a) Neutrons Current Density

If the neutron density is not constant throughout a given volume, neutrons move from regions of higher density to regions of tower density. If we consider a unit area in a volume occupied by a neutron gas. There will be an excess of neutrons passing through this area in one direction as compared to the number of neutrons crossing it in the opposite direction. When the neutron density becomes uniform in the neighbourhood of our unit area, the net flow stops. The net number of neutrons travelling per second through the unit area which is normal to the direction of flow is known as *neutron current density.* It is a vector quantity and can be started in cartesian coordinates as :

$$J = J_x i + J_y j + J_x k. \quad ...(2)$$

According to Fick, neutron current density J and the space rate of change of the neutron density dn/ds are proportional to each other,

$$J = -D\, dn/ds. \quad ...(3)$$

Here D is the *diffusion coefficient* and negative sign indicates that the neutron current is in the direction of decreasing neutron density. In standard vector notation dn/ds = grad n = ∇n = density gradient. For monoenergetic neutrons flux $\phi = nv$, hence

$$J = -\frac{D}{v}\frac{d\phi}{ds} = -D_0 \nabla\phi. \qquad ...(4)$$

Hence D_0 is the diffusion constant. Since grad $\phi = \left(\frac{\partial \phi}{\partial x}\right) i + \left(\frac{\partial \phi}{\partial y}\right) i + \left(\frac{\partial \phi}{\partial z}\right) k$, hence after substituting this value of grad ϕ in eqn. (4) and comparing it with eqn. (2), we have

$$J_x = -D_0 \frac{\partial \phi}{\partial x},\ J_y = -D_0 \frac{\partial \phi}{\partial y} \text{ and } J_x = -D_0 \frac{\partial \phi}{\partial z}. \qquad ...(5)$$

Each of these components represents the net current density along the respective axis, which is the difference between a current density component in the positive direction and a current density component in the opposite direction. Hence

$$J_x = J_x^+ - J_x^-;\ J_v = J_v^+ - J_v^-;\ J_z^+ - J_z^-. \qquad ...(6)$$

It can be shown that current density components are related with neutron flux ϕ and macroscopic scattering cross-section as

$$J_{x_+} = \frac{\phi}{4} - \frac{1}{6\Sigma_s}\frac{\partial \phi}{\partial x},\ J_{x_-} = \frac{\phi}{4} + \frac{1}{6\Sigma_s}\frac{\partial \phi}{\partial x}. \qquad ...(7)$$

$$\therefore \qquad J_x = -\frac{1}{3\Sigma_s}\frac{\partial \phi}{\partial x} = -\frac{\lambda_{tr}}{3}\frac{\partial \phi}{\partial x}. \qquad ...(8)$$

Thus we have $D_0 = 1/2\ \lambda_{tr}$. ...(9)

(b) Neutron Leakage Rate

Let us consider a small volume element dV = dxdydz in a rectangular frame of reference. The neutron leakage rate can now be calculated as the difference between the number of particles entering and the number of particles exiting.

Let us first consider a direction parallel to the x-axis. The number of particles crossing the face dy dz is J_x dy dz. After entering they continue to move in the same direction and at a distance dx they encounter the opposite face of volume element. If $\frac{\partial J_x}{\partial x}$ is the rate of change of the current density per unit length along the x-axis, the total number of particles crossing the opposite face of volume element dV will thus be

$$\left[J_x + \left(\frac{\partial J_x}{\partial x}\right)dx\right]dy\ dz.$$

Hence the net leakage rate of neutrons from the volume element in x-direction $= \left(\frac{\partial J_x}{\partial x}\right)dx\ dy\ dz.$

The same calculations lead to the leakage from the volume element in y- and z-directions. Thus

$$\text{Total leakage rate} = \left(\frac{\partial J_x}{\partial x} + \frac{\partial J_y}{\partial y} + \frac{\partial J_x}{\partial z}\right) dx\ dy\ dz$$

$$= -\ D_0(\partial^2\phi\ \partial x^2 + \partial^2\phi/\partial y^2 + \partial^2\phi/\partial z^2)\ dx\ dy\ dz$$

$$= -\ D_0\ \nabla^2\phi\ dV = -\ D\nabla^2 n\ dV. \qquad ...(10)$$

The ∇^2 is used for the sum of second derivatives of a function and is known as the Laplacian. In spherical and cylindrical coordinates respectively it is given by

$$\nabla^2 = \frac{1}{r^2}\frac{\partial}{\partial r}\left(r^2\frac{\partial}{\partial r}\right) + \frac{1}{r^2\sin^2\theta}\left[\frac{\partial^2}{\partial\phi^2} + \frac{\partial}{\partial\theta}\left(\sin\theta\frac{\partial}{\partial\theta}\right)\right] \qquad ...(11)$$

$$\nabla^2 = \frac{1}{r}\frac{\partial}{\partial r}\left(r\frac{\partial}{\partial r}\right) + \frac{1}{r^2}\frac{\partial^2}{\partial\theta^2} + \frac{\partial^2}{\partial z^2}. \qquad ...(12)$$

(c) Thermal Neutron Diffusion

To simplify the problem, let us treat thermal neutrons as a monoenergetic group. All the neutrons of the group are assumed to have same energy. On the average net energy change remains zero when colliding with the nuclei of the moderator. At a given point r (x, y, z) of the moderator, the neutron density n(r) will depend upon following three factors : (a) The rate of production of thermal neutrons per unit volume, Q; (b) The rate of absorption of thermal neutrons per unit volume, $n v \Sigma_a$; (c) The rate of diffusion or leakage per unit volume, $-\ D\nabla^2 n$.

Hence the rate of change of neutron concentration with time $\partial n/\partial t$ = rate of production–rate of absorption–rate of leakage

$$= Q - n\nu\Sigma_a - \left(-D\nabla^2 n\right) = Q - \frac{n\nu}{\lambda_a} + \frac{1}{3}\lambda_{lr\nu}\nabla^2 n \qquad \text{....(13)}$$

This is the diffusion equation for neutrons that has to be solved for each particular case according to the requirements of the problem. When a steady state has been established the neutron density n at any given point inside the moderator will be independent of time t. This is in fact reached after enough time has elapsed, so that the production rate of neutron and the leakage and absorption have adjusted themselves to an equilibrium value. Since production of thermal neutrons is mainly due to slowing down of fast neutrons to thermal energies, hence Q can be replaced by slowing down density q. Thus *steady state equation* can be written as

$$\nabla^2 n - \frac{3}{\lambda_{tr}\lambda_a} n + \frac{3q}{\nu\lambda_{tr}} = 0. \qquad \text{...(14)}$$

In terms of neutron flux ϕ we can write

$$\nabla^2 \phi - \frac{3}{\lambda_{tr}\lambda_a}\phi + \frac{3q}{\lambda_{tr}} = 0. \qquad \text{...(15)}$$

Let us imagine that the source is a point. We can imagine surrounding the source point a sphere of infinitesimal radius so that in all the space outside it, the q = 0. Thus the diffusion equation for these regions reduces to

$$\nabla^2 \phi - \frac{3}{\lambda_{tr}\lambda_a}\phi = 0 \text{ or } \nabla^2 \phi - \frac{1}{L^2}\phi = 0, \qquad \text{...(16)}$$

where $L^2 = \lambda_{tr}\lambda_a/3$. This term L is known as *thermal diffusion length.* To get a solution of eqn. (16), the ∇^3 is expressed in coordinates most appropriate to the geometry of the problem.

We shall write the diffusion equation for the spherical reference system, making use of the Laplacian in spherical form for the region outside the point source. The equation is hence

$$\frac{1}{r^2}\frac{\partial}{\partial r}\left(r^2\frac{\partial \phi}{\partial r}\right) - \frac{1}{L^2}\phi = 0. \qquad \text{...(17)}$$

After substituting $R = r\phi$, we get

$$\frac{1}{r^2}\frac{d}{dr}\left(r\frac{dR}{dr} - R\right) - \frac{1}{L^2}\frac{R}{r} = 0, \text{ or } \frac{d^2R}{dr^2} - \frac{R}{L^2} = 0.$$

This equation has the solution

$$R = Ae^{r/L} + Be^{-r/L} \quad ...(18)$$

or
$$\phi = \frac{A}{r}e^{r/L} + \frac{B}{r}e^{-r/L}. \quad ...(19)$$

These arbitrary constants A and B, can be determined from the boundary conditions. Since the neutron flux is finite for all values of the variable r, hence the constant A must be zero. To determine B we can make use of the physical condition that the total neutron absorption rate for the medium must be equal to the source strength Q. Therefore

$$Q = \int_0^\infty \phi \Sigma_a \, dV = \int_0^\infty \phi \Sigma_a \, 4\pi r^2 dr$$

$$= \int_0^\infty \frac{B}{r} e^{-r/L} \Sigma_a \, 4\pi r^2 dr = 4\pi \Sigma_a \, BL^2$$

or
$$B = \frac{Q}{4\pi \Sigma_a L^2} = \frac{3Q}{4\pi \lambda_{tr}} \quad ...(20)$$

∴
$$\phi = \frac{3Q}{4\pi\,\lambda_{tr}} \frac{1}{r} e^{-r/L}. \quad ...(21)$$

Similarly if the *neutron source is in the form of an infinite plane* that emits Q thermal neutrons per unit area per second. We shall consider a slab of material infinite in the y- and z-directions but of finite thickness in the x-direction. Thus the steady state equation becomes

$$\frac{d^2\phi}{dx^2} - \frac{1}{L^2}\phi = 0. \quad ...(22)$$

The general solution of this equation can be written as

$$\phi = Ae^{x/L} + Be^{-x/L}.$$

Again A = 0 if ϕ is to remain finite for all values of x. B can be evaluated, we can make use of the condition that the neutron current density J_x parallel to the source plane is

$$J_x = -\frac{1}{3}\lambda_{tr}\frac{d\phi}{dx} = \frac{1}{3}\lambda_{tr}\frac{B}{L}. \quad ...(23)$$

Since the neutron source emits Q neutrons per sec per unit area in both +ve and –ve directions, hence by virtue of symmetry of the conditions on both sides of the source one can write

$$J_x = \frac{1}{2}Q = \frac{1}{3}\lambda_{tr}\frac{B}{L} \quad \text{or} \quad B = \frac{3QL}{2\lambda_{tr}}.$$

and the solution thus comes out to be

$$\phi(x) = \frac{3QL}{2\lambda_{tr}} e^{-x/L} \quad \text{...(24)}$$

$$\text{If } x = L, \ \phi = \frac{3QL}{2\lambda_{tr}} \frac{1}{e} \quad \text{...(25)}$$

Thus for an infinite plane as neutron source, L is the distance from the source at which the neutron flux is reduced by a factor 1/e and is equal to the *relaxation length* which is more generally known as the *thermal diffusion length.* This is a measure of the air line distance travelled by neutron between the point of its origin as a thermal neutron A and the point of its absorption B.

For a point source, a relationship between the mean square distance from the source to the point of neutron absorption $\overline{r^2}$ and L^2 can be established by the following procedure. Let the flux of neutrons at a distance r from the point source emitting neutrons at a given rate be ϕ neutrons per unit area per second. If Σ_a is the macroscopic absorption cross section, the rate of absorption of neutrons will be $\Sigma_a \phi$ neutrons per unit volume per second. Thus the number of neutrons absorbed per second in a spherical shell of thickness dr at a distance r from the neutron source is $4\pi r^2 dr \ \Sigma_a \phi$. This is the probability of absorption of neutron when it is within the spherical shell of thickness dr. Therefore the mean square distance travelled by a neutron from the source to the point where it is absorbed is defined by

$$r^2 = \frac{\int_0^\infty r^2 4\pi r^2 dr \, \Sigma_a \phi}{\int_0^\infty 4\pi r^2 dr \, \Sigma_a \phi} = \frac{6L^4}{L^2} = 6L^2$$

or $$L^2 = \frac{1}{6}\overline{r^2} \quad \text{...(26)}$$

In other words, the square of the diffusion length is equal to one sixth the average of the square of the shortest distance that a neutron travels from the point where the neutron is emitted to where it is absorbed. The parameter L^2 is called the *diffusion area.*

Similarly we have an expression for the slowing down length $L_s^2 = \frac{1}{6}\overline{r_s^2}$, where r_s is the distance measured in a straight line from the point where a fission neutron is produced to the point where it becomes a thermal neutron.

(d) The Exponential Pile

It is used for measuring diffusion length for a moderator experimentally. It is a rectangular column built from the moderator material with the neutron source at one end. The neutron flux in such a column falls off exponentially as $\phi \sim e^{-x/L_1}$ where L_1 is a combination of the diffusion length L for the moderator and the geometry of the column and can be given as

$$\frac{1}{L_1^2} = \frac{1}{L^2} + \frac{\pi^2}{a^2} + \frac{\pi^2}{b^2} \quad ...(27)$$

Here a and b are the linear dimensions of the base of the column.

(e) Fast Neutron Diffusion and Fermi Age Equation

Let us now consider the diffusion of neutrons during the pre-thermal or slowing down stage. By virtue of their slowing down, neutrons undergo considerable energy changes while diffusing.

The neutrons change their average energy and diffuse until they reach thermal energy. In the slowing down region neutron density per energy interval n(E) depends on the difference between the slowing down density q(E+dE] into the energy interval dE and the slowing down density q(E) out of it. This difference is zero for thermalized neutrons. In order to simplify the calculations, a continuous loss of energy for a slowing down neutron is assumed instead of actual discontinuous energy loss.

Let us investigate the neutron balance for a moderator in which there is no neutron absorption, $\Sigma_a = 0$. and in which neutrons are not being produced but diffuse through it before getting thermalized.

If n is the initial number of neutrons per unit volume with energies between E and E + Δ E, the only physical processes that can cause neutron n to change are : *(i) diffusion of neutron into or out of the unit volume and (ii) slowing down of neutron into the energy interval ΔE and out of it.* The neutron flow and neutron balance in a space energy. Neutron leakage by diffusion along the space co-ordinate is compensated by a neutron excess flowing along the energy co-ordinate into the energy interval dE.

In terms of neutron density the rate of diffusion is given by

$$-D\nabla^2 n = -\frac{1}{3}\lambda_{tr}\, v \nabla^2 n. \quad ...(28)$$

The number of neutrons slowing down into the energy interval dE and remaining will be equal to the difference of the inflow and the outflow, that is

$$q(E+dE)-q(E)-\left(\frac{\partial q}{\partial E}\right)dE \qquad ...(29)$$

Hence $$-\frac{1}{3}\lambda_{trv}\nabla^2 n=\left(\frac{\partial q}{\partial E}\right)dE. \qquad ...(30)$$

By successive differentiation

$$\nabla^2 q = E\xi v\Sigma_s\nabla^2 n \text{ or } \nabla^2 n=\frac{\nabla^2 q}{E\xi v\Sigma s}. \qquad ...(31)$$

Substituting this value of $\nabla^2 n$ in equation (30). we get

$$\nabla^2 q=\frac{\partial q}{-\left(\frac{\lambda_S\lambda_{tr}}{2\xi E}\right)dE}\cdot \qquad ...(32)$$

Let us introduce a new variable τ, such that

$$d\tau=-\frac{\lambda_s\lambda_{yt}}{3\xi}\frac{\partial E}{E} \qquad ...(33)$$

or $$\tau=\int_{E_0}^{E} d_\tau=\int_{E}^{E_0}\frac{\lambda_S\lambda_{tr}}{3\xi}\frac{\partial E}{E}=\frac{\lambda_{tr}\lambda_S}{3\xi}\log\frac{E_0}{E}. \qquad ...(34)$$

Substituting this new variable in equation (32), we have

$$\nabla^2 q - \partial q/\partial\tau = 0. \qquad ...(35)$$

This is the general equation of diffusion including all energies and is known as the *Fermi age equation* and the variable τ as the *Fermi age* or as the *neutron age*. Despite its name it must be emphasized that dimensions of τ are not of time as can readily be seen from equation (35), but are those of $(\text{length})^2$. It plays the same role in the slowing down equation as does the time in the heat conduction equation and is more strictly analogous to the variable kt/cp in heat conduction.

As $(1/\xi)$ log E_0/E represents the average number of collisions a neutron, undergoes with the moderator nuclei to change energy from E_0 to E. If C is used for this number, equation (34) becomes

$$\tau=\frac{1}{3}\lambda_{tr}\lambda_S\,C=\frac{1}{8}\lambda_{tr}\,\Lambda_S. \qquad ...(36)$$

where As represents the total zig-zag path length of a neutron between the moment of the beginning of its slowing down and the moment of its arrival at energy E, which is usually thermal energy. Here we see that Λs is quite analogous to λ_a in relation for thermal diffusion length. We can, therefore, define a quantity τ_0 such that

$$\tau_0 = \frac{1}{3}\lambda_{tr}\,\Lambda S = L_f^2. \qquad ...(37)$$

The quantity L_f defined in this manner is called the *fast diffusion length.* For a point source emitting fast neutrons the solution of the Fermi age equation comes out to be

$$q(r) \propto e^{-r^2/4t/t^{3/2}} \qquad ...(38)$$

This is a neutron distribution for a given τ. Its form is similar to the Gaussian distribution. At the origin (r = 0) it has its maximum value. The maximum is higher, the smaller the value of τ. K is clear from equation (34) that the value of τ is determined by the choice of the lower limit for to energy in the integral. The more advanced the degree of federation of the neutron, the lower its energy and the greater the corresponding value of τ. Thus we see that *the neutron age is a direct measure of the degree of moderation.* The slowing down density distribution represented that the *thermal neutrons have the larger range and a large fermi age where as the faster neutrons have smaller range and small fermi age.* The root mean square distance for a given neutron age can be obtained by the similar procedure adopted for thermal neutrons, as

$$\overline{r^2} = \int_0^{\infty} r^2(q)4\pi^2\,dr \Big/ \int_0^{\infty} q(r)4\pi r^2\,dr$$

$$= 6\tau. \qquad ...(39)$$

If $\tau = \tau_0$, the neutron age that corresponds to the termination of the moderation process and the attainment of thermal energies of the neutron, then we have

$$\overline{r^2} = 6\tau_0 = 6L_f^2. \qquad ...(40)$$

Thus the neutron age τ is defined as the one sixth of the mean square distance from the point of creation to the point where their energy has been reduced to a value E which corresponds to that τ. This is analogous to that of L^2 in thermal diffusion. The sum of τ_0 and L^2 is called the *migration area.*

$$\tau = -\int_t^{t_0} \frac{\lambda_S \lambda_{tr}}{3\xi} \frac{dE}{E} = -\int_t^{t_0} \frac{\lambda_{trv}}{3} dt = -\int_t^{t_0} D dt$$

$$= -(t_0 - t) \cdot \frac{1}{(t_0 - t)} \int_t^{t_0} D dt = -(t_0 - t)\overline{D} = \overline{D} \times (t - t_0)$$

= Average diffusion coefficient × Slowing down time. ...(41)

Experimental Determination of Fermi's Age : A neutron source, for example polonium beryllium mixture emitting 10^{11} neutrons/m^3 sec approximately, is placed inside a pile of graphite. Indium foils are placed in cadmium boxes so as to exclude thermal neutrons. Because of the strong resonance of indium of 1.46, the values of saturation activity I_∞ are proportional to slowing down density q at this energy. Hence equation (38) becomes

$$I_\infty = Ke^{-r^2/4\tau} \quad ...(42)$$

where K is a constant. Logarithm of this gives

$$\log I_\infty = \log K - \frac{r^2}{4\pi} \quad ...(43)$$

A plot I_∞ verses r^2 then given a straight line whose negative slope is $1/4\tau$. Thus fermi age can be determined.

SOLVED EXAMPLES

Example 1:

Calculate the neutron age and slowing down length for fission of neutrons 2.0 MeV average energy to thermal energy 0.025 eV in graphite and beryllium. Given : Σ_s = 33.5 for graphite and 57 for beryllium.

Solution:

Equation (79) can be written as

$$\tau_0 = \frac{\lambda_{tr}\lambda_s}{3} \frac{\log\left(\frac{E_0}{E}\right)}{\xi} = \frac{\lambda_s}{\left(1 - \frac{2}{3A}\right)} \frac{\lambda_s}{3} \frac{\log\left(\frac{E_0}{E}\right)}{\frac{2}{\left(A + \frac{2}{3}\right)}}$$

$$= \frac{1}{6\Sigma_s^2} \frac{A+\frac{2}{3}}{1-\frac{2}{3A}} \log\left(\frac{E_0}{E}\right).$$

For graphite, A = 12

$$\tau_0 = \frac{1}{6\times(33.5)^2} \frac{12+\frac{2}{3}}{1-\frac{2}{36}} \log\frac{2\times10^6}{0.025}$$

$$= 3.65 \times 10^{-2} m^2 = 365 \ cm^2.$$

$\therefore \quad L_f = 0.192$ m.

Similarly for beryllium, A = 9

$\tau_0 = 9.7 \times 10^{-3}$ and $L_f = 9.8 \times 10^{-2}$ m.

Example 2:

Calculate diffusion length for thermal neutrons in graphite. Given: σ_a = 3.2 milli barns, σ_s = 4.8 barns, and ρ = 1.62 gm/cm².

Solution:

Diffusion length L is given by the relation

$$L = \left[\frac{\lambda_{a\Lambda tr}}{3}\right]^{1/2},$$

where $$\lambda_a = \frac{1}{\Sigma_a} = \frac{1}{N_0\sigma_a} = \frac{A}{N\rho\sigma_a}$$

$$= 12/6.02 \times 10^{26} \times 1.62 \times 10^3 \times 3.2 \times 10^{-31}$$

$$= 38.40 \text{ m}.$$

$$\lambda_{tr} = \frac{\lambda_s}{\frac{1-2}{3A}} = \frac{\frac{1}{\Sigma_s}}{\frac{1-2}{3A}} = \frac{\frac{A}{N\rho\sigma_s}}{\frac{1-2}{A}}$$

$$= \frac{12/6.02\times10^{16}\times1.62\times10^3\times4.8\times10^{-28}}{\frac{1-2}{3\times12}} = 0.0254 \text{ m}$$

$\therefore \qquad L = 0.572$ m.

Example 3:

Calculate the energy of the neutrons reflected from the crystal of lattice 2.3 Å, if the glancing angle of the incident neutrons for Ist order reflection is 3.5°.

Solution:

$$\text{Energy of the neutrons } E = \frac{n^2h^2}{8md^2 \sin^2 \theta}$$

$$= \frac{\left(6.6256\times10^{-27}\right)}{8\times1.6748\times10^{-24}\times\left(2.3\times10^{-8}\right)^2\left(0.061\right)^2} = 1.04 \text{ eV}.$$

Example 4:

Calculate the average energy loss for neutrons that have made one collision with carbon nuclei using ξ = 0.158 for carbon.

Solution:

Since $\qquad \xi = \overline{\Delta \log E} = \overline{\log\left(E_0/E\right)}$, hence

$$\left(\overline{\frac{E_0}{E}}\right) = e^{\xi} \text{ and } \left(\overline{\frac{E_0 - E}{E_0}}\right) = 1 - e^{\xi} = 0.1462.$$

Another method of finding average fractional energy loss is the use of eqn. (21). Then, we get

$$\left(\overline{\Delta E/E_0}\right) = \frac{1}{2}(1-\alpha) = \frac{1}{2}\left[1 - \frac{(A-1)^2}{(A+1)^2}\right].$$

This result shows that it is incorrect to use ξ which is the mean value of the log and not the log of the mean value for finding average energy loss per collision.

Example 5:

Calculate the fractional energy loss of a neutron being scattered through an angle of 60° in the C.M. system in collision with 8.3. Hence *prove that the maximum fractional energy loss in this is greater than that for collision with U^{238} nuclei.*

Solution:

Relation (11) can be written in terms of α, as

$$\frac{E_1}{E_0} = \frac{1}{2}(1+\alpha) + \frac{1}{2}(1-\alpha)\cos\phi.$$

$$\text{As } \alpha = \frac{(A-1)^2}{(A+1)^2} = \left(\frac{8}{10}\right)^2 = 0.64$$

$$\therefore \qquad \frac{E_1}{E_0} = \frac{1}{2}(1+0.64) + \frac{1}{2}(1-0.64)\cos 60 = 0.91$$

$$\text{Hence} \qquad \frac{\Delta E}{E_0} = 1 - \frac{E_1}{E_0} = 0.09 = 9\%.$$

$$\text{As} \quad \left(\frac{\Delta E}{E_0}\right)_{max} = 1 - \frac{(A-1)^2}{(A+1)^2}, \text{ hence}$$

$$\text{for Bc}^2 \left(\frac{\Delta E}{E_0}\right)_{max} = 1 - \left(\frac{8}{10}\right)^2 = 0.36 \text{ or } 36\%.$$

$$\text{and for} \qquad U^{238} = -\left(\frac{237}{239}\right)^2 = 0.02 \text{ or } 2\%.$$

Example 6:

A manganese foil of 100 mg is placed in an exponential pile and exposed to the thermal neutron flux for 2 hour. Now activity of the foil is found to be 200 disintegrations/sec. If the cross-section of manganese for neutrons of 0.025 eV is 13.2 barns and the half life of the radioactive isotope is 2.58 hour.

Calculate the thermal neutron flux of the pile at the point of insertion of the foil, if the point is assumed at temperature 50°C. Calculate the diffusion length L or thermal neutrons for the material if the base is a rectangular of 1.5 m × 2m and the neutron flux is found to be twice when the experiment is repeated with the foil inserted at a distance of 16 cm. nearer the base of the pile.

Solution:

Neutron energy corresponding to a pile temperature of 50°C

$E = kT = 1.38 \times 10^{-23} \times 323 = 4.46 \times 10^{-21}$ Joule

$= 0.028$ eV

Absorption cross-section

$$\sigma_a = \sigma_0 \left(\frac{E_0}{E}\right)^{1/2} = 13.2\left(\frac{0.025}{0.028}\right)^{1/2} = 12.5 \text{ barns}$$

Activity $\quad A = \lambda N_0(1 - e^{-\lambda_t}) = \Sigma_a \phi V(1 - e^{-\lambda_t})$.

Here $\quad \lambda = \frac{0.693}{T_{1/2}} = \frac{0.693}{2.58} \text{ hr} = 0.269 \text{ hr}^{-1}$.

$\therefore \quad (1 - e^{-\lambda_t}) = 1 - e^{-0.538} = 0.416$.

Total number of Mn^{55} nuclei in foil

$$N = 6.02 \times 10^{26} \times 0.1 \times 10^{-3}/55 = 1.094 \times 10^{21}$$

$\therefore$ Neutron flux $\quad \phi = 200/1.094 \times 10^{21} \times 12.5 \times 10^{-28} \times 0.416$

$= 3.5 \times 10^8$ neutrons/m² sec.

As the neutron flux $\phi \propto e^{-x/L}$

$$\therefore \quad \frac{\phi_1}{\phi_2} = \frac{e^{-x_1/L_2}}{e^{-x_1/L_1}} \text{ or } \log_e\left(\frac{\phi_2}{\phi_1}\right) = \frac{x_2 - x_1}{L_1}$$

$$\text{or} \quad \frac{1}{L_1} = \log_e\left(\frac{\phi_2}{\phi_1}\right)(x_2 - x_1) = \frac{0.693}{0.16} = 4.33 \text{ m}^{-1}.$$

From eqn. (72), we get

$$\frac{1}{L^2} = \left(\frac{1}{4.33}\right)^2 + \left(\frac{3.14}{1.5}\right)^2 + \left(\frac{3.14}{2}\right)^2 = 11.9$$

$\therefore \quad L = 0.29$ metre.

Example 7:

How many collisions are required for neutrons to loss, on the average, 90% of an initial energy of 2 MeV in a Be^9 moderated assembly ($\xi = 0.209$)? Compare this with the total member of collisions required to reach thermal energies.

Solution:

If neutrons of energy E_1 are slowed down to energy E_2, then number of collisions required to reduce this amount of energy

$$n = \frac{1}{\xi}\log\left(\frac{E_1}{E_2}\right) = \frac{1}{0.209}\log_e \frac{100}{1} = 22.$$

Similarly the number of collisions required to reduce the energy to thermal energy (0.025 eV) is

$$n = \frac{0}{0.209}\log_e\left(\frac{2\times10^6}{0.025}\right) = 87.$$

Example 8:

Calculate the average distance travelled by a neutron away from an infinite planer neutron source in a moderator before being absorbed. Compare it with the root mean square of the distance travelled.

Solution:

The average distance is obtained by averaging over all distances

$$\therefore \qquad x = \frac{\int_0^\infty xe^{-X/L}dx}{\int_0^\infty e^{-x/L}dx = L}$$

$$= \text{Thermal diffusion length.}$$

Similarly we can obtain

$$\overline{x^2} = \frac{\int_0^\infty x^3e^{-x/L}dx}{\int_0^3 e^{-x/L}dx = 2L^3}.$$

or $$x_{n.m.s} = \sqrt{(2)}L = \sqrt{(2)}\,\overline{x}.$$

EXERCISES

1. A certain (1/ν) absorber reduces the intensity of a thermalized neutron beam of effective temperature T = 300°K by 10%. What would be the reduction in intensity if the same experiment were done at a temperature of 327°K.

2. Calculate the age difference for neutrons slowing down in graphite between neutrons that have reduced the indium resonance energy of 1.44 eV and neutrons that have reduced thermal energies 0.025 eV.

3. Calculate the thermal diffusion length for neutrons in an aqueous solution of neutral uranium salt (weight composition 5% uranium, 95% of light water.

4. Compute the average logarithmic energy loss per collision, the number of collisions needed to reduce the energy from 2 MeV to 0.025 eV, the slowing down power and the moderating ratio of bismuth, given σ = 30 millibarn and σ_s = 9 barn.

5. Calculate the maximum energy of neutrons released by a (γ, n) reaction from a phosphorus target by 17 MeV γ-rays.

6. A crystal spectrometer has a beryllium crystal with a lattice spacing of 0.7323 Å. What are the Bragg angles for the first order reflection and of neutrons of energy 1, 3, 5 and 30 eV respectively?

5

Nuclear Reactors

NUCLEAR CHAIN REACTION

The particular processes with which we are concerned are nuclear fission reactions in which fresh neutrons arc supplied by the fission processes. Thus to maintain a chain reaction, the neutrons must be produced at a constant rate and number of neutrons released per fission, v, must be sufficiently greater than 1 to compensate neutron loss. Since v is constant for a given fissionable material and, therefore, beyond human control. The only alternative is to reduce the neutron loss. The achievement of a chain reaction with uranium depends on a favourable balance among following six processes:

(1) Neutron may escape without being captured.

(2) Neutron may be absorbed by other materials present in the assembly without causing fission.

(3) Neutron may be absorbed by U^{235} without causing fission.

(4) Neutron may be absorbed by U^{235} nucleus causing fission.

(5) Neutron may be absorbed by U^{238} without producing fission.

(6) Neutron may be absorbed by U^{238} and may cause a fission when its energy is greater than the threshold energy.

Thus we see that processes (1) and (3) create new neutrons while other remaining processes remove available neutrons from the assembly. *A device for releasing nuclear energy through fission at a controlled rate is called nuclear reactor or chain reacting pile.* In nuclear weapons, uncontrolled nuclear reactions are used as most rapid and largest energy released is desired. The neutron balance in a thermal reactor can be described in terms of a cycle that shows what happens to the neutrons. Consider this cycle starting with the fission of U^{235} nucleus by a thermal neutron, v fast neutrons are emitted in this process. Some of them, having energy greater than the fission threshold for U^{238}, can cause fissions in U^{238} with a consequent increase in the number of fast neutrons. The total number of fast neutrons from fission is thus increased from v to v∈, where ε is greater than one and is called the *fast fission factor*. In certain reactors ∈ = 1.03.

From a finite size reactor core, some fast neutrons may leak out before being slowed down to thermal energies. To account for this, we introduced l_f the leakage factor for fast neutrons and find that the number of fast neutrons remaining in the core long enough for slow down is vε $(1 - l_f)$. These neutrons are slowed-down by collisions with moderator atoms, but during the slowing down process some of them may be captured by U^{238} to form U^{239} which decays to Np^{239} and then to Pu^{239}, particularly at energies in the region of strong resonances. We introduce the resonance escape probability p and find that out of v∈ $(1 - l_f)$ neutions, a fraction vε $(1 - l_f)$ p escapes while the fraction v∈ $(1 - l_f)$ (1 – p) is captured and goes to form Pu^{239}. The neutrons which escape are slowed down to thermal energies. Some of these thermal neutrons may leak out of the core. To account for this we introduce l_{th}, the *leakage factor for thermal neutrons* and find that the number of slow or thermal neutrons remaining in the core is vε $(1 - l_f)$ p $(1 - l_{th})$. A fraction of this remaining number is absorbed by uranium and the remaining other part by other materials. The fraction of thermal neutrons absorbed by the fuel as compared to all thermal neutron absorptions in the assembly is called the *thermal utilization factor f*. If we introduce this term we obtain the number of thermal neutrons that are actually absorbed by the fuel, which is v∈ $(1 - l_f)$ p $(1 - l_{th})$f.

Not all of the neutrons absorbed in the fuel caue of U^{238} nuclei, some are absorbed by U^{238} to form Pu^{239} while others are absorbed m U^{235} to form U^{236}. If σ_a is the cross-section of thermal neutrons absorbed in uranium and σ_f that of neutrons cause fission of U^{235}. Thus the final number of second generation neutrons absorbed in the fuel is v∈ $(1 - l_f)$ p $(1 - l_{th})$ $f\sigma_f/\sigma_a$. This quantity is known as *reproduction factor or multiplication factor* and is denoted by k. The quantity v (σ_f/σ_a) represents the number of fast fission neutrons, produced by each thermal neutrons, absorbed by the fuel and is called η. This quantity can be expressed as

$$\eta = v \frac{N_{235}\,\sigma_f(U^{235})}{N_{235}\,\sigma_f(U^{235}) + N_{235}\,\sigma_r(U^{235}) + N_{238}\,\sigma_o(U^{238})} \qquad ...(1)$$

where σ_r (U^{235}) stands for the cross-section for irradiative capture by U^{235}. For natural uranium $N_{235}/N_{238} = 0.00715$. It has been found that v = 2.5. The experimental values for the microscopic cross-section are σ_f (U^{235}) = 549 barns, σ_r (U^{235}) barns, σ_a (U^{238}) = 2.8 barns, hence η = 1.32.

$$\therefore \quad k\, \eta \in (1 = l_f)\, p\, (1 - l_{th}) f \qquad ...(2)$$

If the core is of very large size, the leakage of fast as well as slow neutrons can be disregarded. The multiplication factor is then denoted by kw and is given by

$$k_\infty = \eta \varepsilon p f \qquad ...(3)$$

This relation is known as the *four-factor formula.*

In an infinite system the condition of the self-sustaining chain reaction is that each neutron generation just replaces the previous one *i.e.*, k = 1. The system is then said to be critical. Fora finite system, because in each neutron generation some neutrons are loset from the sides of the system, the *critical condition* is that k must be greater than unity, the exact amount by which it has to be greater than unity depending upon the shape of the system and the arrangement of the materials in the system, etc. If k is greater than unity for the infinite system, the number of neutrons will increase steadily. This is known as a *super-critical* state. On the other hand for k less than unity the number of neutrons would decrease to zero and this state is known as *sub-critical* state.

In the special case of a reactor in which the fuel contains U^{236} only and no U^{238}, both the factors e and p are practically unity. The formula for kw thus reduces to

$$k_\infty = \eta f \qquad ...(4)$$

For an assembly of finite dimensions, the effective multiplication factor keff will be less than k_∞ by a factor $(1 - l_f)\,(1 - l_{th})$. Thus

$$k_{eff} = k_\infty\, (1 - l_f)\, (1 - l_{th}) \qquad ...(5)$$

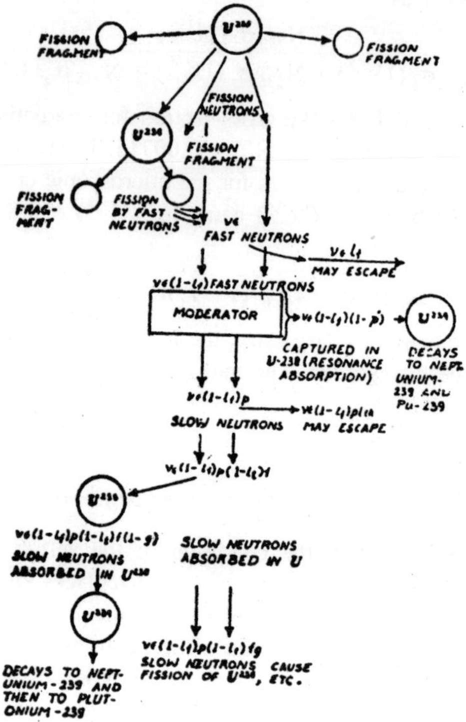

Fig 5.1 : Neutron cycle.

The magnitude of keff determines the speed with which the number of neutrons builds up and the rate at which fission occurs. In a nuclear bomb type of assembly, this build up must take place very rapidly, ($k_{eff} > 1$), whereas in industrial and research reactors this self-multiplication must be slow enough to allow the fission rate to remain always under the control of the operator ($k_{eff} \simeq 1$). If $k_{ef} > 1$ for an assembly, we can decrease it by progressively reducing the reactor size, there by increasing the neutron loss through leakage from the assembly. The reactor size of the assembly for which $k_{eff} = 1$ is called the critical size.

THE CRITICAL SIZE OF A REACTOR

The size of a reactor that operates under the condition of exact balance between neutron production and neutron loss is known as critical *size and the reactor itself is said to be critical.* For such a reactor, a relation between its geometric properties and the material properties of the assembly is known as the critical equation. For all types of reactors, both homogeneous and heterogeneous, we start with eqn.

$$\nabla^2 \phi - \frac{3}{\lambda_{tr}\lambda_a}\phi + \frac{3q}{\lambda_{tr}} = 0 \qquad ...(6)$$

where q is the slowing down density which is a function of the space co-ordinates r and the neutron age τ. Thus the Fermi age equation can be written as

$$\nabla^2 q\,(r,\tau) = \frac{\partial q(r,\tau)}{\partial \tau} \qquad ...(7)$$

The solution of this equation can be written in the form of a product of two functions R (r) and T (τ). We can thus write

$$q(r, \tau) = R(r)\, T\,(\tau). \qquad ...(8)$$

By taking the second derivatives with respect to the space coordinates and first derivative with respect to τ only and substituting in equation (7), we get

$$T\,(\tau)\, \nabla^2 R = R(r)\, \partial T/\partial \tau$$

or
$$\frac{\nabla^2 R}{R(r)} = \frac{1}{T(\tau)}\frac{\partial T}{\partial \tau} = -\,B^2\,(\text{say}) \qquad ...(9)$$

where $-B_2$ is a constant. Thus we have two differential equations.

$$\frac{1}{T(\tau)}\frac{\partial T}{\partial \tau} = -\,B^2 \text{ or } \frac{\partial T}{\partial \tau} + B^2\, T(\tau) = 0 \qquad ...(10)$$

and
$$\frac{\nabla^2 R}{R(r)} = -\,B^2 \text{ or } \nabla^2 R + B^2 R(r) = 0 \qquad ...(11)$$

The solution of eqn (10) is

$$T\,(\tau) = T_0 e^{-B^2 r} \qquad ...(12)$$

where T_0 is the value of T initially when $\tau = 0$.

At the beginning of the slowing down process,

$$q\,(\tau = 0) = q_0 = R(r)\, T(0) = R(r)\, T_0 \qquad ...(13)$$

$$\therefore \quad q = R(r)\, T_0 e^{B2} = q_0 e^{-B^2 r} \qquad ...(14)$$

We can also express q_0 in terms of the physical properties of the assembly. Number of neutrons per unit volume per sec. available for slowing down is (given by the rate of production of fission neutrons and is equal to the neutron multiplication factor $\in f\eta$ per thermal neutron absorbed. The rate of thermal neutron absorption per unit volume of the reactor is $\phi\Sigma_a$. Hence

Initial slowing down density = rate of fast neutrons available

or $$q_0 = \phi(r)\,\Sigma_a \varepsilon f\eta \quad ...(15)$$

Thus equation (14) becomes

$$q = \phi(r)\,\Sigma_a \varepsilon f\eta\, e^{-B^2 r} \quad ...(16)$$

If we consider neutron absorption by the moderator, then the corrected eqn. obtained by multiplying the solution by p is

$$q = \phi(r)\,\Sigma_a \varepsilon f\eta p\, e^{-B^2 r}$$
$$= k_\infty \Sigma_a \phi e(r)\, e^{-B^2 r} \quad ...(17)$$

Substituting this value of q in eqn. (6), we get

$$\nabla^2\phi - \frac{3}{\lambda_{tr}\lambda_a}\phi + \frac{3k_\infty e^{-B^2 r}}{\lambda_{tr}\lambda_a}\phi = 0$$

or using $\frac{1}{3}\lambda_{tr}\lambda_a = L^2$, we get

$$\nabla^2\phi + \left(\frac{k_\infty e^{-Bh^2 r - 1}}{L}\right)\phi = 0 \quad ...(18)$$

This is a general *diffusion eqn. for an actual reactor.*

Reactor Buckling

Using eqns. (7) and (17), we get

$$k_\infty \Sigma_a e^{-B^2 r}(\nabla^2\phi + B^2\phi) = 0,$$

or $$\nabla^2\phi + B^2\phi = 0 \quad ...(19)$$

Since the thermal neutron flux distribution ϕ (r) depends on the geometry of the assembly, the value of B^2 is similarly determined by the reactor geometry and is known as geometrical buckling, denoted by B_g^2. Combining eqns (18) and (19), we get

$$\left[\frac{k_\infty}{(1 + L^2B^2)}\right] e^{-B^2 r} = 1 \qquad ...(20)$$

This is the critical equation. The left hand side is also denoted by k_{eff}. The numerical value of B^2 depends upon the physical properties of the reactor material and is, therefore, called the *material buckling* ($B_m^{\ 2}$) of the reactor. Geometrical buckling $B_g^{\ 2}$ decreases as we increase geometrical dimensions of a critical reactor. But this increase results in a k_{eff} greater than unity. As material buckling $B_m^{\ 2}$ does not change with the reactor size. Hence for a super critical reactor ($k_{eff} > 1$), $B_m^{\ 2}$ must be greater than $B_g^{\ 2}$ Similarly, a reduction in size makes the reactor subcritical ($k_{eff} < 1$) for which $B_g^{\ 2}$ is greater than $B_m^{\ 2}$.

Nonleakage Factors

Fraction of fast neutrons that does not leak out of the assembly during slowing down, l_f, is given by the ratio of the actual production rate of thermal neutrons over the maximum possible for a reactor of infinite size. Thus

$$l_f = \frac{k_\infty \Sigma_a \phi e^{-B^2 r}}{k_\infty \Sigma_a \phi} = e^{-B^2 r} \qquad ...(21)$$

Thermal non-leakage factor l_{th} = 1—Thermal leakage factor

$$= 1 - \frac{\text{Thernal diffusion rate}}{\text{thermal diffusion rate} + \text{Thermal absorption rate}}$$

$$= 1 - \frac{-\frac{1}{3}\lambda_{tr}\nabla^2\phi}{-\frac{1}{3}\lambda_{tr}\nabla^2\phi + \Sigma_a\phi} = - \frac{\Sigma_a\phi}{-\frac{1}{3}\lambda_{tr}\nabla^2\phi + \Sigma_a\phi}$$

$$= \frac{1}{1 + \frac{1}{3}\lambda_{tr}\lambda_a B} = \frac{1}{1 + B^2L^2} \qquad ...(22)$$

$$\therefore \quad k_{eff} = k_\infty\, l_{th} l_f = k_\infty\, e^{-B^2 r}/(1 + L^2B^2)$$

$$= \frac{k_\infty}{(1 + B^2\tau)(1 + L^2B^2)} = \frac{k_\infty}{1 + B^2(L^2 + \tau)} = \frac{k_\infty}{1 + B^2M^2} \qquad ...(23)$$

Hence $L^2 + \tau$ is replaced by M^2, the migration area. M is known as migration length. Thus for large thermal critical reactor

$$k_{eff} = k_\infty/(1 + B^2M^2) = 1 \qquad ...(24)$$

The shape of nuclear reactors are almost exclusively either a rectangular parallelepiped, a sphere, or a cylinder. The most suitable co-ordinate systems are cartesian for the first, spherical for the second and cylindrical for the last. We shall summarised here the solutions for these three co-ordinate systems.

Rectangular Parallelepiped of Sides a, b, c.

The equation of neutron flux in this reactor in cartesian co-ordinates is given by

$$\frac{\partial^2\phi}{\partial x^2} + \frac{\partial^2\phi}{\partial y^2} + \frac{\partial^2\phi}{\partial z^2} + B^2\phi = 0 \qquad ...(25)$$

The solution, the neutron flux, will be the function of x, y, z and can be written as the product of three functions as :

$$\phi = X(x)\, Y(y)\, Z(z). \qquad ...(26)$$

$$\therefore \quad \frac{1}{X}\frac{\partial^2 X}{\partial x^2} + \frac{1}{Y}\frac{\partial^2 Y}{\partial y^2} + \frac{1}{Z}\frac{\partial^2 Z}{\partial z^2} + B^2 = 0 \qquad ...(27)$$

Let us assume that

$$\frac{1}{X}\frac{\partial^2 X}{\partial x^2} = -\alpha^2, \ \frac{1}{Y}\frac{\partial^2 Y}{\partial y^2} = -\beta^2 \text{ and } \frac{1}{Z}\frac{\partial^2 Z}{\partial z^2} = \gamma^2, \qquad ...(28)$$

$$\therefore \quad \alpha^2 + \beta^2 + \gamma^2 = B^2 \qquad ...(30)$$

The solution of equation $\left(\frac{1}{X}\right)\frac{\partial^2 X}{\partial x^2} = -\alpha^2$ is supposed to be

$$X = A_1 \cos \alpha x + A_1' \sin \alpha x \qquad ...(30)$$

Because sine function is not a symmetric function, hence the arbitrary constant A_1' has to vanish. Using now the boundry condition $\phi = 0$ for $x = \alpha/2$, we get $A_1 \cos a\, \alpha/2 = 0$ or $\alpha = \pi/a$.

Thus the solution becomes

$$X = A_1 \cos (\pi/a)\, x \qquad ...(31)$$

Similarly we have $\quad Y = A_2 \cos (\pi/b)\, y \qquad ...(32)$

and $$Z = A_3 \cos (\pi/c)\, z \qquad ...(33)$$

$$\phi = A \cos (\pi/a)x \cos (\pi/b)y \cos (\pi/c)z \qquad ...(34)$$

and $$B^2 = (\pi/a)^2 + (\pi/b)^2 + (\pi/c)^2. \qquad ...(35)$$

It can be shown that the minimum critical volume of a parallelepiped is a cube (a = b = c) for which $B^2 = 3\pi^2/a^2$.

$$\therefore \qquad V_{min} = a^3 = 3^{3/2} + \pi^3/B^2 = 161/B^3 \qquad ...(36)$$

Spherical Reactor of Radios R

Because of spherical symmetry, if we take the origin of the reference system at the centre of the sphere, the solution will be independent of θ and ϕ. The diffusion equation thus can be written as

$$\frac{\partial^2\phi}{\partial r^2} + \frac{2}{r}\frac{\partial\phi}{\partial r} + B^2\phi = 0 \qquad ...(37)$$

The solution of this equation is

$$\phi = (A_1/r) \sin Br + (A_2/r) \cos Br. \qquad ...(38)$$

With the use of the boundary conditions : ϕ is finite at r = 0 and is zero at r = B, we get $A_2 = 0$ and BR = π or R = π/B. Thus the solution for the spherical reactor is

$$\phi = (A_1/r) \sin (\pi/R)r. \qquad ...(39)$$

Under the critical condition the minimum volume

$$V_{min} = \frac{4}{2}\pi R^2 = \frac{4}{3}\pi\left(\frac{\pi}{B}\right)^2 = \frac{130}{B^2} \qquad ...(40)$$

Cylindrical Reactor

The diffusion equation in cylindrical co-ordinates with cylindrical symmetry (term containing derivative of angular co-ordinates θ = 0) is

$$\frac{\partial^2\phi}{\partial r^2} + \frac{1}{r}\frac{\partial^2\phi}{\partial r} + \frac{\partial^2\phi}{\partial z^2} + B^2\phi = 0 \qquad ...(41)$$

The solution of this equation can be written as the product of two functions as

$$\phi = R(r)\, Z(z) \qquad ...(42)$$

Thus equation (41) can be written as

$$\frac{1}{R}\frac{\partial^2 R}{\partial r^2} + \frac{1}{rR}\frac{\partial R}{\partial r} + \frac{1}{Z}\frac{\partial^2 Z}{\partial z^2} + B^2 = 0 \qquad ...(43)$$

Let $$\left(\frac{1}{Z}\right)\frac{\partial^2 Z}{\partial z^2} = -\alpha^2 \qquad ...(44)$$

The solution is $Z = A_2 \cos \alpha z + A_2' \sin \alpha z$.

To have cylindrical symmetry A_2' has to vanish because since function is not a symmetric function. We then have

$$Z = A_2 \cos \alpha z \qquad ...(45)$$

By using boundary condition $\phi = 0$, for $z = H/2$, we have

$$A_1 \cos \alpha H/2 = 0$$

$$\therefore \qquad \frac{\alpha H}{2} = \frac{\pi}{2} \quad \text{or} \quad \alpha = \frac{\pi}{H} \qquad ...(46)$$

Hence $$Z = A_2 \cos (\pi/H/)z. \qquad ...(47)$$

Substituting the value of a and hence that of term containing Z in equation (43), we have

$$\frac{d^2R}{dr^2} + \frac{1}{r}\frac{dR}{dr} + \left(B^2 - \frac{\pi^2}{H^2}\right)R = 0 \qquad ...(48)$$

Writing $B^2 = \frac{\pi^2}{H^2} = K^2$ and multiplying by r^3, we have

$$r^2 = \frac{d^2R}{dr^2} + r\frac{dR}{dr} + K^2r^2R = 0$$

Calling $rK = \mu$, we can write finally

$$\mu^2 = \frac{d^2R}{d\mu^2} + \mu\frac{dR}{d\mu} + \mu^2 R = 0 \qquad ...(49)$$

It is a linear differential equation of second order in cylindrical co-ordinates and known as the Bessel's eqn. of zero order. Its solution is the Bessel functions of zero order and can be written as

$$R = A_3 J_0(\mu) \qquad ...(50)$$

Combining the results we have finally

$$\phi = AJ_0 (Kr) \cos (\pi/H)z. \qquad ...(51)$$

From the Bessel function tables of zero order we find that $J_0(Kr)$ reaches the value 0 when Kr = 2.405. It corresponds to r = radius R of the cylinder. Hence K = 2.405/R.

Substituting this value of K in equation (51), we have

$$\phi = AJ_0 \left[\left(\frac{2.405}{R}\right) r\right] \cos\left(\frac{\pi}{H}\right) z \qquad ...(52)$$

$$\therefore \qquad B^2 = \frac{\pi^2}{H^2 + (2.405/R)^2} \qquad ...(53)$$

For the smallest critical volume of the cylinder V = ($\pi R^2 H$) should be minimum. Hence

$$\frac{dV}{dH} = \frac{d}{dH}\left(\pi \times \frac{H^2 (2.405)^2}{B^2 H^2 - \pi^2} \times H\right) = 0$$

or
$$B^2 = \frac{2\pi^2}{H^2} \qquad ...(54)$$

Therefore,
$$V_{min} = \frac{\pi (2.405)^2}{2\pi^2} \left(\frac{\sqrt{3}\pi}{B}\right)^3 = \frac{148}{B^3} \qquad ...(55)$$

From the above results it is apparent that for a nuclear reactor using the same materials the most convenient shape is the spherical one as this shape requires the minimum volume. It is also clear from above result that the minimum volume of any of the shapes considered is proportional to the inverse of tie coefficient B^3. The quantity B depends upon the average number of neutrons created per fission, the diffusion length and the slowing down length. As these depend upon reactor material, the quantity B is known as the *material buckling* and denoted by B_m. Hence the minimum volume is defined as the critical condition under which geometrical buckling is equal to the material buckling. This determines the amount of material required for the reactor to become critical.

The Effect of a Reflector

The critical size of a reactor core can be reduced by surrounding it with good moderating material. The latter has the properties of being a good neutron scatterer and does not usually absorb neutrons to any great extent. This material acts as a reflector. Graphite is often used for this purpose. The reactor with a reflector has certain distinct advantantages over the bare reactor which are stated as follows :

(a) Improved Neutron Economy

The reflector knocks neutrons which leave the reactor back into it and thus reduces neutron leakage from the core. The reflector also acts as a moderator for the fast neutrons that have entered it from the core. Since the absence of neutron absorbing material in the reflector reduces neutron loss due to resonance absorption, the moderation of fast neutrons in the reflector will be more efficient that in the core itself.

(b) Possibility of Fuel Saving

For constant core size reacton the improvement in the neutron economy reduces the amount of fuel or the fuel concentration required to achieve criticality. It means that reflector increases fuel saving. We also know that the reflector decreases the leakage or neutrons from the reactor and thus decreases the critical volume. Hence it is possible to reduce the size of the nuclear reactor including the reflector to below the size of a similar barer eactor with the sums fuel concentration.

(c) Improved Reactor Power Utilization

Since the power production rate is proportional to the average neutron flux, the reactor can be operated at a higher total power output for the same maximum neutron flux. The improvement in the power utilization is a consequence of the flattening of flux across the reactor core occuring due to the presence of the reflector. By virtue of this (flux flattening) effect the power production is also move uniform over the core volume.

GENERALASPECT OF REACTOR DESIGN

In spite of numerous possible variations in the design and components of reactor systems, there are number of general features which all reactors possess in common to a greater or lesser extent Let us consider and compare few features in the given order.

1. Fuel

The material containing the fissile isotope is called the reactor fuel. The fuels that can be used are uranium containing U^{235} in its natural concentration of 0.715% or in an enriched proportion. Th^{232} and U^{235}, which can be converted partly at least into fissile material. The form in which a fuel is used in a reactor depends upon various circumstances. Solid fuel elements are used in most of the reactors. Some reactors have been designed which make use of fluid fuels.

2. Moderators and Reflectors

Materials used to reduce the neutron energy, known an moderators, are graphite, light water, heavy water, beryllium and its oxide and possibly certain organic compounds. As previously noted, the good moderators are usually materials of low mass number having small absorption cross-section and large slowing down power. The ordinary water has the greatest *sdp* but an appreciable neutron absorption cross-section. It is attractive as a moderator because of its low cost. Heavy water is the best available material for neutron moderation. It has a very small absorption cross-section and the highest moderating ratio of all moderators. Carbon in the form of graphite has been employed as moderator or reflector in a number of reactors. It has a very small absorption cross-section and the second highest moderating ratio of all moderators. Both beryllium and its oxide are good moderators having small absorption cross-section and the best sdp of all metals. Metallic beryllium is much more expensive than other moderators. Hydrogeneous organic compounds have proved to be suitable as moderators.

Above mentioned materials arc not suitable for use as the reflector for the reactor in which fast neutrons are used. To avoid this situation the reflector must be a material of high mass number.

3. Reactor Coolants

The materials employed, to remove the heat that is being generated in the reactor core as a result of fissioing taking place, are known as *coolants*. These materials are circulated through the core tor the purpose of abstracting the heat and transferring it to the outside of the core. An ideal coolant should have as little effect on the neutron as possible, ideally it should not absorb nor moderate the neutrons, should not react chemically with the other materials which it contacts at the system, should not break up under irradiation, should acquire intense long lived radioactivity during its passages through the reactor, should have a low vapour pressure at the operating temperature of the reactor, should be reasonably easy to handle, should be able to remove large amounts of heat for a small expenditure of pumping power and should not be costly. Actually no one coolant meets all these requirements, a final choice for any particular reactor is a compromise based on these requirements. The materials proposed for coolants are ordinary water, heavy water, liquid metals such as sodium or sodium potassium alloy or mercury, organic liquids and gases. Each type has its advantages and disadvantages.

At ordinary temperatures both heavy and light water are good coolants. For reactors operating at high temperatures, high pressures are required to prevent boiling. Liquid metals have been proposed for use at higher temperatures. The best of these is sodium. It does not require pressurization even at very high temperatures. On the other hand, it is very reactive with water and oxygen and becomes radioactive due to neutron capture when passed through reactor. As a compromise between water and liquid sodium, certain organic compounds e.g. polyphenyls, were suggested. However the heat removal properties of these hydrocarbons are inferior to those of water or liquid sodium. In several of the earliest reactors air was used as a coolant. It is not satisfactory at high temperatures because it reacts chemically with so many materials. Carbon dioxide is also employed in large power reactors. Since it is used as coolant in graphite-moderated reactors. The chemical reaction with graphite is a problem at temperatures greater than 500°C. By the addition of some carbon-monoxide or methane it can be used upto 600°C. Actually helium is a good coolant but it is very costly.

4. Control Materials

In thermal reactors control is achieved by means of a neutron absorbing materials. The material should not become radioactive as a result of neutron capture. Became of its availability and ease of fabrication cadmium was used in early reactors. It can be utilized at low temperature work only because it has low melting point. An alloy of silver with 15 per cent of indium and 5 percent of cadmium has been also employed in some reactors. This alloy has a higher melting point than Cd and large cross-section for neutron capture also over a wider energy range. The most common neutron absorber used for reactor control is boron. This element has very high melting point and large cross-sections for neutron absorption- Boron is often used incorporation with stainless steel, aluminium or carbon.

The control elements are commonly located in the core in the form of either rods or plates, but in some reactors it is more convenient to have the control elements in the reflector close to the core. In a thermal reactor the control rods are moved in to decrease the fission rate or neutron flux and out to increase it. In fast reactors materials cannot be used because the cross-sections are small. In this case control is achieved by adjusting neutron escape by moving part of the reflector or some of the fuel elements.

5. Reactor Shielding

All nuclear reactors, except those operating at very low powers, are sources of intense neutron and γ-radiation and therefore, represent hazard to persons in the immediate vicinity of the reactor. Provisions for their health protection are made by surrounding the reactor core with a radiation shield.

This shield is generally known as the *biological shield* because its primary function is of health protection. It consists generally a layer of concrete, about 6 to 8 ft. thick and is capable of absorbing both γ-rays and neutrons. In reactors operating at high powers, that part of the shield which is in immediate contact with the core heats up considerably and requires special cooling facilities in order to prevent it from cracking or suffering other heat damage. Shield required for this purpose is known as thermal shield. It is fairly close to the core and consists of a few inches of iron or steel.

CLASSIFICATION OF REACTORS

Reactors may be classified in a wide variety of ways. For example they may be classified by the manner in which the fuel and the moderator are mixed. In a *homogeneous reactor* the fuel and moderator arc mixed in the form of a solution. **Most of the nuclear** reactors are of the heterogeneous type, in which fuel is concentrated in plates, rods, or hollow cylinders, which are distributed in a regular pattern (lattice) within the moderator.

Reactors are further classified according to the energy of the neutrons that cause fission. These reactors ire known as *thermal, intermediate* or fast according to the thermal neutrons (0.025 eV), intermediate neutrons (1 to 10 keV) and fast neutrons (above than 10 keV) respectively. Reactors are also classified according to their purpose. The types to be examined here are (a) research reactors, (b) production reactors, and (c) power reactors.

(a) Research Reactors

Research reactors are used primarily to supply neutrons for physical research and radioisotope manufacture. In these reactors the total energy liberated is comparatively small, since they operate at a low level of reactivity. In these cases cooling is required only to prevent over heating. We shall describe here five main types of research reactors :

1. Graphite-Moderated. Research Reactor

The graphite-moderated, heterogeneous, natural uranium reactor has the distinction of being the first reactor built and operated. This first reactor, now called CP-1 (Chicago Pile-1), was put into operation in 1942. This was operated without coolant flow at a power level of 200 watts, Similar assemblies with air as coolant, like the GLEEP (graphite low energy experimental pile) and BEPO (British experimental pile) at Harwell (England) and the X—10 Clinton pile at Oak Ridge and BNL (Brookhaven National Laboratory) are well known examples as they have been the work horses of reactor technology and research. All these reactors use fuel in the shape of cylindrical rods, or variations on this shape, clad usually in aluminium. "Several hundred tons of graphite are used as moderator and reflector, with several hundred cylindrical channels passing right through the graphite to allow the introduction and positioning of the fuel elements. Cooling gas is made to pass over the elements through the channels at low pressure. The power and neutron flux for different reactors are given in Table 1.

Table 1 : Graphite—Moderated, Air-coolant Heterogeneous, Research Reactor

Name	Size	Fuel and its arrangement	Reflector	Power (kW) Neutron Flux
CP-1	9'×9'×12'	Uranium (6 tons) its oxide (40 tons) spherical lumbs $\left(D = 2\frac{1}{4}\right)$ at $8\frac{1}{4}$ cubical spacing	1 ft.	0.2
CP-2	18'×19'×20'	Uranium (10 tons) its oxide (40 tons) similar to CP-1	"	0.2–2 3×10^3

GLEEP	Cylinder D = 9.5' H = 17'	Uranium (33 tons) slugs D=0.97", H=12", $7\frac{1}{4}$ spacing, Al-clad	2ft.	100 3.7×10^{10}
BEPO	Cylinder D = 20 H=20'	Uranium (40 tons) slugs (D=0.9", H=12", $7\frac{1}{4}$ spacing. Al-clad	3ft.	4000-6000 10^{12}
X-10	20' Cube	Uranium (49 tons) slugs (Dù1.K//SS4" spacing Al-clad.	2ft.	1000-3800 $\sim 10^{12}$
BNL	25' Cube	Uranium (100 tons) slue (D=1.1", H=4"), 8" spacing Al-clad	4,5'	30000 4×10^{12}

2. Water Boiler Type Reactor

Perhaps one of the roost misleading names of a reactor type is the water boiler. The water boiler is usually a homogeneous mixture of a highly enriched uranium salt dissolved in ordinary water. The name water boiler often leads to a mistaken belief that these reactors are the boiling water type of reactor. Actually the boiling water reactors arc quite diffe- rent and are under the bead "power reactors".

The name arises from the fact that a sudden increase in power will cause the formation of steam bubbles in the solution, which in turn will shut down the reactor quickly. Under normal conditions, the solution does not boil as its temperature is kept below 80°C by circulating water through the coils inside of the core vessel. The first reactor of this type was built as Los Alamos in early 1944. It was named LOPO (low power) because of its low power operation. It was improved in December for high power operation and was named HYPO (high power).

It was further improved for superpower, named SUPO (Super power), and is in operation since 1951. Another water boiler of note is the reactor built at the North Carolina State College in 1953. It is the first reactor owned and operated by a University anywhere in the world. Similar assemblies, like WBNS (water boiler neutron source) California and ARR (Armour research reactor) Chicago and KEWB (kinetic experimental water boiler) California are well known examples. Few properties of these reactors are given in Table 2.

Table 2 : Water Boiler, Water Coolant, Water Moderator, Homogeneous Research Reactor

Name	Size	Fuel	Reflector	Power (kW)/ Neutron flux
LOPO	Sphere 1'dia 15 litre	Enriched uranyl sulphate solution; 15 litres; U^{238} (580 gm.), U^{238} (3380 gm.).	BeO and Graphite	5×10^{-5}
HYPO	"	Enriched U-nitrate solution, U^{235} (897 gm.) U^{238} (5341 gm.).	BeO and Graphite	63×11^{11}
SUPO	"	90% enriched U-nitrate, 0.8 kg. in 13 litres.	Graphite	35-45 1.7×10^{12}
WBNS	"	90% enriched uranyl nitrate solution, U^{235} (700 gms.).	Graphite	0.005 2×10^{5}
North Carolina	Cylinder $(D=10\frac{3}{4}$, H=11')	90% enriched UO_2 SO_4 solution,. U^{235} (848 gm.) in 14 litres.	Graphite	10 5×10^{11}
ARR	15 litre spherical core	88% enriched uranyl sulphate.	Graphite	50 10^{12}
KEWB	Sphere (1,1/4, dia)	90% enriched uranyl sulphate, U^{235} (1750 gm.).	Graphite	50 9×10^{11}

The water boiler reactor has proved to be versatile research tool providing intense sources of neutrons and γ-rays. It has a great merit of simplicity of design and construction. Its use is somewhat limited because of its low power rating.

3. Swimming Pool Reactor

The turn swimming pool reactor is the popular name of light water moderated heterogeneous reactor which consists of a lattice of enriched uranium fuel immersed in a large pool of water. The water in the pool plays the role of moderator, coolant and shield. The chief advantages of this type of reactor are simplicity, low cost, safety, flexibility and accessibility. The first swimming pool reactor was completed in 1950 at the Oak Ridge National Laboratory. It was named BSR because was designed with the view of allowing radiation effects on bulky materials to be observed by immersing them in the water surrounding the core.

The Indian Reactor **Apsara** belongs to this type. This reactor was completed on August 4, 1956 and was the first reactor to go into operation in Asia. It was named Apsars on the name of Legendary sea mermaid because it is half under the tea and half outside. It was designed and built by the Indian engineers and scientists under the guidance of the late Dr. H.J. Bhaba. This reactor is designed for a maximum power level of 1 mw and contains large quantity of light water in a concrete pool. The fuel assembly is attached to steel frames and is suspended from above from an adjustable bridge placed on rails. This bridge spanes the pool. The fuel elements are arranged in the rectangular assembly. These are made of enriched uranium alloy and arc held inside aluminium cans. Few control rods of Cd metal are also introduced in the fuel assembly and are adjusted automatically so that reactor does not become super critical. Holes are provided through graphite reflector through which thermal neutrons are available. When the reactor becomes critical the water surrounding the core emits a brilliant Cerenkov light glow, due to the passage of energetic electrons through water, can be seen from above the pool in the broad day light as well.

Table 3 : Swimming Pool Water Coolant, Water Moderator, Heterogeneous Research Reactor

Name	Size	Fuel	Reflector	Power (kW) Neutron flux
BSR	20'×20'×40'	MTR type 3.5 kg. U^{235}, Fuel core 1'×1'×2'	H_2O or BeO	100-1000 2×10^{13}
NRLR (Naval)	$49' \times 32' \times 28\frac{1}{2}$	MTR type, Fuel core 21"×21"×24'	H_2O or Graphite	100-1000 10^{12}
Baitelle RR		MTR type 2.8 kg. U^{235}	H_2O or Graphite	1000 1.4×10^{13}
Swiss Reactor	10' dia 2' deep	23 MTR type fuel assemblies.	H_20	10-100 10^{11}
Melusine (France)		Enriched uranium	H_2O	1000 10^{13}
Apsara (India)		Enriched uranium alloy	Graphite	100-1000 10^{12}

The first fuel charge was removed from the reactor in Dec. 1964 since it had attached its maximum burn up. Maintenance of its various of fuel elements, obtained from U.K., in July 1965. The standard element of second fuel charge contains 140 gm of U^{235} (80% enriched uranium). For normal operation the reactor core contains, about 25 fuel elements- This reactor has been used for the production of isotopes, irradiation of biological samples, and for testing the various types of reactors and health physics instruments. It has been used for fundamental research in nuclear physics and structure of materials. A large number of isotopes such as P^{32}, S^{35}, I^{131} and Au^{198}, produced in this reactor, are used in research instruments and hospitals. There are about two dozen swimming pool reactors available in the world. Few properties of some reactors are given in Table 3.3.

4. Light-Water-Moderator, Tank type Reactor

This reactor is very similar to pool reactor, in principle, except that the core is suspended in a deep cylindrical tank of water and cannot be moved and it preferred over 'one of the pool type when a large number of holes's for neutron beams in many different directions is an important requirement. A reactor with increased neutrons flux was designed and built in 1952 at the National Reactor Testing Station in Arco, Idaho and was named Material Testing Reactor (MTR).

Table 4 : Light water moderator. Water Coolant, Heterogeneous, Tank Type Reactor.

Name	Size	Fuel	Reflector	Power (kW) Neutron flux
MTR	Overall (32'×34'×40') reactor tank (56' dia, 30' high), fuel assembly (16"×28"×24")	Sandwich type > 90% enriched fuel elements of U-Al alloy, 19 curved vertical fuel plates in 3×3×24 inches box 4.5 kg. U^{235} in Al tank, 54" dia with side extension to form water well	Be and graphite	4×10^{1} 3.7×10^{14}
ETR	32" dia 36" high	49 MTR type fuel elements	Be and Graphite	1.75×10^{5} 3×10^{11}
WTR	Core (D=28", H=36")	60 MTR type fuel assemblies	H_2O	10^{14}
HFR (Highflux)		90% enriched MTR type fuel elements U^{235} (4.2 kg.)	H_2O	2×10^{14} 4×10^{11}

The MTR fuel elements are heterogeneous plate type elements. These plates, often referred to as sandwich type plates, contain an inner core of uranium-aluminium alloy which Is 0.02" thick. This core is clad (bread) on either side by 0.02" thick aluminium. Eighteen of the plates are brazed together in aluminium side plates with a spacing of 0.117" between plates to form which is generally known as the MTR type element. Each one of the MTR elements contains 140 gm. of U^{235} and the critical mass for the reactor is approximately 3 kg. The properties of few other reactors of this type are listed in the Table 3.4.

5. Heavy-Water-Moderator, Tank type Reactor

The desirability of obtaining heavy water for use as a neutron moderator lies in the extremely small thermal absorption cross section deuterium. Because of this property the neutron economy in a heavy water system is greatly improved. The first heavy water moderator reactor was the Chicago Pile 3 (CP-3) which went into operation on may 1944 at the Argonne National Lab. and was dismantled in January 1950. It is a simple tank type of reactor. Canada has specilized in heavy water natural uranium reactor because of the abundant supply of natural uranium and no facilities for enrich ment of natural uranium. The first reactor constructed in Canada was completed in Sep. 1945.

Cirns : formerly known as CIR, is India's second research reactor, built as a joint Indo-Canadian project under the Colombo Plan at Trombay. This reactor was completed on 10 July, 1960. The name CIR stands for Canadian Indian Reactor and the new name cirus is the name of the star close to our solar system. It is a natural uranium heavy water moderated and light water cooled high flux research reactor with a terminal power of 40 MW. It is Operated round the clock. The standard fuel rod contains about 55 kg. of natural uranium. For normal operation the reactor core contains about 185 fuel rods and 23 tonnes of heavy water. This reactor is used for basic and applied research in physics, chemistry, biology, metallurgy and reactor technology, for the production of large quantities of high activity radio isotopes as well as plutonium-239 and uranium-233.

Zerlina : Is India's third research reactor. It attained criticality on Jan. 14, 1961. The name zerlina stands for *zero energy lattice investigation nuclear assembly* as it was made to study the lattice assemblies of different kinds. In this reactor natural uranium is used as a fuel and heavy water as a moderator. The standard fuel rod contains about 45 kg. of natural uranium. The standard core contains 80 fuel rods and about 8 tonnes of heavy water. It is operated on a two shift basis upto a maximum power of 100W and is being extensively used for physics and studies of uranium—heavy water lattice.

Purnima : It is India's first experimental zero energy fast reactor. It attained criticality on May 22, 1972. It uses plutonium oxide as fuel. It is used for the training of personnel in the operation of plutonium fuelled fast reactors. Experiments were carried out by scientists to measure parameters used in the construction of the core of the pulsed fast reactor at Kalpakham.

R-5 : It is a thermal research reactor with a nominal power of 100 MW and is located adjacent to 40 MW cirus reactor.

The natural uranium fuel elements of cirus and zerlina and the plutonium oxide for purnima are fabricated in Trombay. This fabrication facility went into operation in June 1959. The present capacity of this plant is about 30 to 44 tones of fuel elements per year.

(b) Production Reactors

The purpose of a production reactor to convert fertile into fissile material. The fissible materials are used as a fuel in other reactors. In Hanford works production reactors, the natural uranium is used as a fuel, graphite as a moderator and water as a coolant. Because of the relatively large number of neutrons captured by a hydrogen nuclei present in water, the proportion available for capture by U^{238} is reduced considerably. The efficiency of conversion of U^{238} into Pu^{239} is thus appreciably less than unity. The ratio is always smaller than 0.8 in this type of reactor.

The efficiency of the conversion process can be increased by decreasing resonance escape probability. It is mainly by capture of neutrons in the resonance region. U^{238} is converted into Pu^{239}. Since maximum value of multiplication factor in uranium graphite reactor is 1.07, hence any decrease in the resonance escape probability can make the reactor impossible to attain criticality. Multiplication factor would be a large as 1.25 if heavy water is used as a moderator. A substantial decrease in the resonance escape probability is thus permissible without making reactor impossible to attain criticality. Thus we see that conversion efficiency of a production reactor using heavy water as a moderator is greater than the reactor using graphite as a moderator.

(c) Power Reactors

The primary purpose of a power reactor is the utilization of the fission energy produced in the reactor core and to convert it into useful power. Heat generated in fission process is used, either directly, or indirectly, to produce steam at high temperature and pressure or to heat a compressed gas. The steam or heated gas is used as a working substance in a turbine. The turbine can be connected to an electric generator or can be used as a source of mechanical power. Choice of the heat exchange material is governed by three important considerations.

(d) It must have a high thermal conductivity so as to carry away the heat efficiently and give it up in a heat exchanger.

(b) It must have a low neutron capture cross-section so as not to upset the reactor characteristic.

(c) It must not be decomposed by intense radiation.

Now we shall discuss some of the reactor types which appear to show the greatest promise for power production.

1. Pressurized Water Reactor

This is a heterogeneous reactor that uses slightly enriched uranium (1.4 to 2% of U^{235}) as fuel and light water as a moderator and coolant. This type of reactor has received much attention in the United Slates. The fuel elements are in the form of rods or plates. The core is contained in a vessel under a pressure of 1000 to 2000 pounds per square inch. The pressurized water is circulated through the reactor core, from which it removes heat, and then through on external heat exchanger. Heat exchanger is used to abstract heat energy from the coolant in order to make it available for conversion into other forms of energy. High pressure is maintained in the core in order to raise the boiling point of the coolant to high enough temperature only because a consequence of the laws of thermodynamics is that heat can be more coefficiently convened into useful energy if it is released at a high temperature.

This first full scale nuclear power plant in the United States at Shipping port, Pennsylvania, was completed at the end of 1957. The first one produced originally 60 mega watts of electric power. This power was latter increased to 100 megawatts with a capability of going to 150 megawatts of electricity Several other reactors of this type have been (or are being) constructed in the United States and in other countries e.g., France, Italy and U.S.S.R.

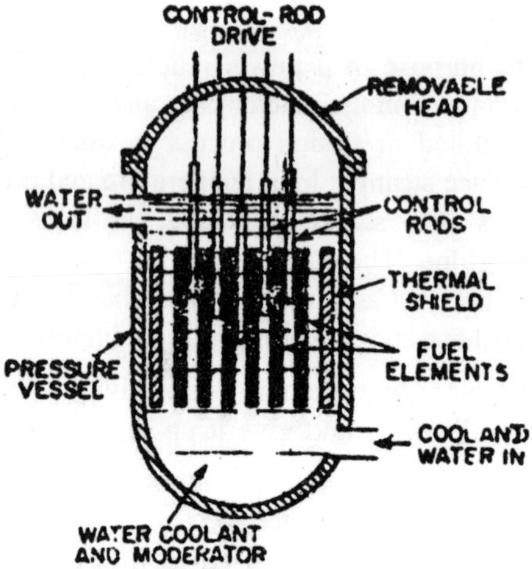

Fig. 5.2 : Schematic diagram of the pressure vessel and core of a pressurized water reactor

2. Boiling water Reactor

The boiling water reactor is a small nuclear power plant designed to allow steam to be generated directly in the reactor core. This uses light water as moderator coolant. This reactor system is more advantageous than pressurized water reactor. The pressure in the reactor vessel does not have to be much greater than the required steam pressure while in PWR the pressures used to prevent boiling are significantly larger. External heat exchangers are also not required. Enriched uranium is used as a fuel.

The experimental boiling water reactor at Argonne National Laboratory was the first reactor of this type. It was designed in the end of the year 1956 and electricity at the power of 5 megawatts was produced. The first boiling water reactor to produce substantial amount of electric power was started up in 1960 at the nuclear power station, near Chicago, Illinois. Central station power plants utilizing boiling water reactors are also in operation in other parts of U.S.A. as well as in Italy and Germany. The maximum power produced by most recent plants is 1000 megawatts of electricity and the steam produced is at a temperature of 285° and a pressure of about 1000 psi. The 420 MW energy station, at Tarapur (India), consists of two enriched uranium reactors of the boiling water type. These reactors were built with the help of the Central Electric Company of the U.S.A. under a turn key contract, on the basis of specifications provided by the Tarapur Atomic Power Project of the Department of Atomic Energy. Nuclear power began to flow from the station when it became operational at 8.15 p.m. on April I, 1969. These reactors operated extremely well in the first few years. However, some problems were raised in the years thereafter. Now since 1975 the performance is rather good.

This type of reactors are particularly adapted to integral super-heating. The steam is produced in one part of the core and super-heated in another part. The first such plant, completed in 1964, was the Boiling Nuclear Superheat (BONUS) reactor in Puerto Rico. In this reactor water is boiled in the central region of the core and the steam is superheated in the outer region. The second superheat reactor, completed in 1965, is at the Pathfinder power station at Sioux Falls. Sout Dakota. The arrangement is reverse of that in the first one. The temperature of exist steam is 480° in the first case and 440°C in the second one.

3. Heavy Water Moderated Reactor

There arc two advantages of reactors moderated by heavy water as compared with ordinary water as a moderator. First, the total cost of the core and cost of the consumed fuel arc both less in the heavy water moderated reactor because natural uranium or very slightly enriched material is used as a fuel. Second, the conversion ratio for the regeneration of fissile material can be close to unity. Both factors decrease the cost of the power, produced. As the high price of the heavy water increases the cost of power, this type of reactor cannot be used for large plants with an electrical capacity of 500 MW.

The first proto-type of a heavy water moderated power reactor is the Nuclear Power Demonstration Reactor (NPDR) completed in 1962 at Canada. It is of the pressurized water reactor type with heavy water as moderator and coolant instead of the light water. The coolant leaves the reactor at 277° and produced steam at about 230° in the heat exchanger.

The first boiling water reactor which uses heavy water as moderator is the Halden (Norway) reactor. In this reactor heavy water coolant-moderator boils and produces heavy steam. It is only one of its kinds in existence and has a designed thermal powers of 20 megawatts.

The second nuclear power station (430 MW) near. Kola at Rana Pratap Sagar in Rajasthan employs heavy water moderated reactors using natural uranium as fuel. The first unit of the reactor became critical in 1972 and started feeding power in 1973 into the power system of Rajasthan. In this power station the design and bulk of the equipment have come from Canada, but the construction and installation activities were done by the Department of Atomic Energy. Unfortunately there have been failures in two stages due to deficiencies in design and manufacture of the turbine generator; designed in U. K. and supplied from Canada. For the second unit of this station under construction, a major step in indigenisation has been taken. The reactor vessel, shields, boilors and fuelling machines have been made in and India and the responsibility has been taken over by Indian Scientists and Engineers.

The third nuclear power station (470 MW) is under construction at Kalpakham about 100 km away from Madras. In this instance complete responsibility for design. Engineering and construction of the project rests with the Department of Atomic Energy. The design generally follows the design used at Rajsthan, significant changes have been made in increase the output of the units by about 10%. This is 80% Indian, only 20% components are from aboard. Apart from reactor vessel and shields, boilers and fuelling machines, the equipment manufactured in India include the turbine generators, zirconium alloy components and the entire instrumentation and control system. The total fuel charge as well as heavy water will come from Indian plants.

The fourth Indian Atomic power plant is under construction at Narora. It will have two atomic reactors of 235 MW each and well deliver 440 MW of power to the northern power grid when completed. It will use natural uranium at fuel and heavy water as moderator like the power stations at Ranapratap Bagh and Kalpakham. This power station will be 100% from Indian resources.

4. Organic Moderated Reactor

Another method for achieving high temperatures at moderate pressures in a thermal reactor is to moderase and cool with an organic liquid containing a high hydrogen content and of his high boiling point. The organic moderated Reactor Testing Experiment (OMRE), was completed in 1957 at the National reactor Testing Station. Arco, Idaho. The organic liquid used in this reactor is a mixture of di and ter-phenyle and has a normal boiling point at atmospheric pressure of about 340°C and 370°C at the operating pressure of some 200 psi. The temperature is limited to a maximum of 325°C because the thermal decomposition of the organic compounds become more than can be tolerated. The experience gained was utilized in the design of a small power plant, called the Piqua Nuclear Power Facility (PNPF) at Piqua Ohia. The coolant used is a mixture of the three terphenyls ($C_{18}H_{14}$) at a pressure of 120 psi. It leaves the reactor at a temperature of 302°1C. Superheated steam at 288°C and a pressure of 440 psi is produced in the heat exchanger. The electric output of this plant is 11.5 megawatts.

5. Gas-Cooled Reactor

Several of the early experimental reactors in the United States were cooled with forced-air. Until 1957, this type of reactors were not developed in the United States because of various reasons. In the mean time air was employed as the coolant in the graphite moderated production reactors at Windscale in U. K. The graphite moderated power reactor with nitrogen carbon dioxide as a coolant and natural uranium as the fuel have been constructed in England, France and other countries. The carbon dioxide gas at a pressure of about 150 to nearly 400 psi in the later designs is pumped through the reactor leaving at a temperature around 400°C. The superheated steam is generated, at temperatures in the approximate range of 320° to 390° in the heat exchanger About 300 megawatts of electricity is produced by most of the larger plants. Helium gas also used in some of the gas cooled reactors. In High Temperature Gas Cooled Reactor, completed in 1966 at Peach Bottom, Pennsylvania, the helium gas leaves the core temperature of 750°C and produces superheated steam at 538°C at and 1450 psi pressure.

The use of beryllium oxide as a moderator in gas cooled power reactor was proposed by F. Daniels and his associates in 1944. With improvements in fabrication Experimental Beryllium Oxide Reactor was constructed but the work was terminated in 1966 because cost of fabrication is a significant disadvantage in power reactor with this moderator.

6. Sodium Graphite Reactor

The high boiling point of liquid sodium makes it an excellent cooled for a high temperature reactor. High temperatures can be achieved without high pressures. Sodium Reactor Experiment was developed to its critical state in 1957. In this reactor graphite was used as moderator, sodium as coolant and either natural or slightly enriched uranium as the fuel.

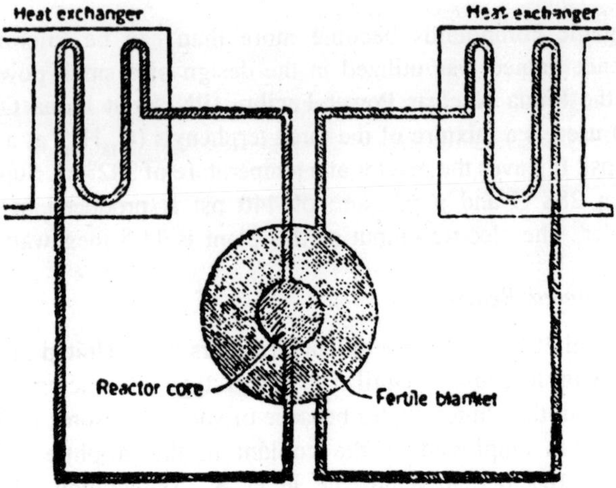

Fig. 5.3 : Schematic Diagram of homogeneous reactor.

Another reactor based upon SRE was completed in 1962. This is known as Hallam Nuclear Power Facility (HNPF). The temperature of the sodium leaving the reactor was 507°C and the steam was generated at a temperature of 446°C and a pressure of 850 psi. Its design power was 76 megawatts of electricity. It is very advantageous because sodium has excellent heat transfer characteristics and does not require excessive pumping power.

SOLVED EXAMPLES

Example 1:

The world's first water boiler reactor, the LOPO nuclear reactor, used an enriched uranium sulphate solution made up as follows:

Element	*weight*	*σ_a(barn)*	*σ_s(barn)*
U^{235}	*580 gm*	*698*	*10*
U^{238}	*3378 gm*	*2.75*	*8.3*
S	*534 gm*	*0.49*	*LI*
0	*14065 gm*	*0.0002*	*4.2*
H	*1573 gm*	*0.33*	*20*

Calculate k_∞ for this reactor.

Solution:

Thermal utilization factor

$$f = \frac{1}{1 + \frac{N_{(238)}\sigma_{a(238)}}{N_{(235)}\sigma_{a(235)}} + \frac{N_{(s)}\sigma_{a(s)}}{N_{(235)}\sigma_{a(235)}} + \frac{N_{(0)}\sigma_{a(0)}}{N_{(235)}\sigma_{a(235)}} + \frac{N_{(H)}\sigma_{(H)}}{N_{(235)}\sigma_{a(235)}}}$$

$$f = \frac{1}{\begin{array}{c}1 + 5.75 \times 3.94 \times 10^{-3} \times 6.76 \times 7.15 \times 10^{-4} \\ +356 \times 2.86 \times 10^{-5} + 637 \times 4.73 \times 10^{-4}\end{array}} = 0.754$$

$$\frac{\Sigma_s}{N_0} = \frac{N_{(0)}}{N_{(238)}}\sigma_{s(0)} + \frac{N_{(H)}}{N_{(238)}}\sigma_{s(H)} = \frac{879.25}{14.19}\times 4.2 + \frac{1573}{14.19}\times 20 = 2475 \text{ barns}$$

$$\therefore \quad \int_E^{E_0} (\sigma_a)_{eff} \frac{dE}{E} = 3.85\left(\frac{\Sigma_s}{N_0}\right)^{0.415} = 99 \text{ barns}$$

$$\text{and} \quad \frac{N_0}{\xi\Sigma_s} = \frac{N_0}{\xi_{(0)}\sigma_{s(0)}N_{(0)} + \xi_{(H)}\sigma_{(H)}N_{(H)}}$$

$$= \frac{14.19}{0.120 \times 4.2 \times 879 + 1 \times 120 \times 1573} = \frac{1}{2249}$$

$$\therefore \quad p = e^{-99/2249} = 0.957$$

Example 2:

Calculate k_∞ for an enriched uranium-graphite moderated assembly which contains 400 molecules of moderator per molecule of uranium and a U^{238}/U^{235} ratio of 75. Given: For graphite, $\sigma_a = 0.0032$ barn, $\sigma_s = 4.8$ barns. $\xi = 0.758$; For uranium, $\sigma_f(235) = 590$, $\sigma_a(235) = 698$, $\sigma_a(238) = 2.75$ and $\nu = 2.46$.

Solution:

Let us first calculate absorption cross section $\sigma_a(U)$ and number of fission neutrons per neutron absorbed η_u for the enriched fuel.

$$\sigma_a(U) = [N(235)\,\sigma_a(235) + N(238)\,\sigma_a(238)]/[N(235) + N(238)]$$

$$= \frac{\sigma_a(235) + \sigma_a(238)\,N(238)/N(235)}{1 + N(238)/N(235)} = \frac{698 + 2.75 \times 75}{1 + 75}$$

$$= 11.89 \text{ barns.}$$

$$\eta_u = [N(235)\,\sigma_f(235)\,v]/[N(235)\,\sigma_a(235) + N(238)\,\sigma_a(238)]$$

$$= \frac{v\sigma_f(235)}{\sigma_a(235) + \sigma_a(238)\,N(238)/N(235)} = \frac{2.46 \times 590}{698 + 2.75 \times 75} = 1.60$$

Calculate f and p and then k_∞ as we did, in the last example.

This value of p is considerably greater then the values obtained for natural uranium. Since f is evaluated with respect to U^{235}, the matching value of η to be used now is 2.08. Hence

$$k_\infty = \epsilon\eta\,pf = 1 \times 2.08 \times 0.957 \times 0.754 = 1.5.$$

Example 3:

Calculate the critical volume of a spherical thermal reactor that uses U^{235} as fuel and heavy water as moderator in molecule ratio of $1 : 10^5$.

Solution:

$$\text{Thermal utilization factor } f = \frac{\sigma_a(U)}{\sigma_a(U) + \sigma_a(M)\dfrac{N_m}{N_u}}$$

$$= \frac{698}{698 + 0.00092 \times 10^5} = 0.8948$$

$\therefore\ k_\infty\ \eta f = 2.08 \times 0.8948 = 1.819$

Thermal diffusion length for the mixture

$$L^2 = L_m^{\,2}(1 - f) = 1.70^2\,(1 - 0.8948) = 0.3038\ m^2$$

$$\therefore\quad k_{eff} = \frac{k_\infty e^{-B^2}}{1 + L^2B^2} = \frac{1.819\,e^{-0.125B^2}}{1 + 0.3038\,B^2} = 1$$

or $B^2 = 2.192\ m^2$

$\therefore$ Critical volume $V_C = 120/53 = 37\ m^3$.

Example 4:

What would be the length of she side of a cubical reactor having geometrical buckling 65. If the reactor is spherical, calculate its critical radius.

Solution:

Geometrical buckling $B^2 = \frac{\pi^2}{a^2} + \frac{\pi^2}{b^2} + \frac{\pi^2}{c^2}$

For a cubical reactor a = b = c

$\therefore\ B^2 = \frac{3\pi^2}{a^2}$ or $a^2 = \frac{3\pi^2}{B^2} = 0.46154$

$\therefore$ a = 0.674 m = 67.4 cm

Critical radius of a spherical reactor

$R = \pi/B = 3.141/8 = 0.39$ m = 39 cm.

Example 5:

Calculate k_∞ for a homogereous, natural uranium-heavy water moderated assembly which contain 50 molecules of moderator per molecule of uranium. Assume natural uranium to contain one part U^{235} to 139 parts of U^{238} and use the following constants : For uranium, $\sigma_a(U) = 7.68$ barns, $\sigma_s(U)$ barns; and for D_2O, $\sigma_a = 0.00092$ barn, $\sigma_s = 10.6$ barns; and $\xi = 0.570$.

Solution:

Thermel utilization factor $f = \frac{N_u\sigma(U)}{N_u\sigma_a(U) + N_u\sigma_a(M)}$

$$= \frac{1}{1 + \frac{N_m\,\sigma_a(M)}{N_u\sigma_a(U)}} = \frac{1}{1 + \frac{0.00092}{7.68}} = 0.994$$

Number of U^{238} atoms only per unit volume $N_0 = N_u(139/140)$

$$\therefore\quad \frac{\Sigma_s}{N_0} = \frac{N_m}{N_0}\sigma_s(M) = N_m \cdot \frac{140}{139 N_u} \times \sigma_s(M)$$

$$= \frac{N_m}{N_u} \times \frac{140}{139} \times \sigma_s(M) = 50 \times \frac{140}{139} \times 10.6 = 530 \text{ barns}$$

$$\therefore\quad \int (\sigma_s)_{eff} \frac{dE}{E} = 3.85\left(\frac{\Sigma_s}{N_0}\right)^{0.415} = 60.2 \text{ barns}$$

$\therefore$ Resonance escape probability $P = e^{-60.2/530\times 0.57} = 0.820$

Hence $k_\infty = \epsilon\eta p f = 1 \times 1.34 \times 0.820 \times 0.994 = 1.09$

This assembly is, therefore, not capable of sustaining a chain reaction.

EXERCISES

1. A nuclear reactor has an effective k of 1.03. How many generations will be required to double the number of neutrons? What time will be required for doubling, assuming an initial energy of 1.5 MeV and a mean free path for fission of 8.0 cm?
2. Liquid sodium used as a reactor coolant has a cross-section for thermal activation of 600 mb. The circulating sodium spends about 1/4 of its time in the reactor core, where the thermal flux is 8.6×10^{12} n cm^{-2} sec^{-1}. What is the specific activity of the coolant at equilibrium? How long will be required for the coolant to reach 3/4 of equilibrium activity?
3. A power reactor operating at a power level of 10^5 watts goes prompt critical due a failure of the control rod system. Assume a multiplying factor of 1.09, and that the negative temperature coefficient stopped the reaction in 5 m sec. Estimate the thermal energy generated during the runaway.
4. A critical reactor of cubical shape uses U^{235} and ordinary water in a molar ratio of 1:400. Calculate the critical mass of U^{235} for this reactor.
5. Calculate the thermal utilization factor for an aqueous fuel solution containing 10% by weight, of U^{235}.
6. Calculate p for a homogeneous, natural uranium-BeO assembly with molecular ratio of 1 : 200.
7. Calculate k_∞ for a homogeneous, natural uranium-graphite-moderated assembly with contains 300 molecules of graphite per molecule of uranium. Assume natural uranium to contain one part of U^{235} to 139 parts of U^{235}. Given; For natural Uranium, $\sigma_a = 7.68$ barns, $\sigma_s = 8.3$ barns; For graphite, $\sigma_a = 0.0032$ barn, $\sigma_s = 4.8$ barn, and $\xi = 0.158$.